Disasters in Australia and New Zealand

Scott McKinnon • Margaret Cook
Editors

Disasters in Australia and New Zealand

Historical Approaches to Understanding Catastrophe

palgrave
macmillan

Editors
Scott McKinnon
Australian Centre for Culture,
Environment, Society and Space
University of Wollongong
Wollongong, NSW, Australia

Margaret Cook
School of Social Sciences
University of the Sunshine Coast
Sippy Downs, QLD, Australia

ISBN 978-981-15-4381-4 ISBN 978-981-15-4382-1 (eBook)
https://doi.org/10.1007/978-981-15-4382-1

This Palgrave Macmillan imprint is published by the registered company Springer Nature Singapore Pte Ltd.
The registered company address is: 152 Beach Road, #21-01/04 Gateway East, Singapore 189721, Singapore

Foreword

An Australasian book about the histories of disasters could not be more timely. Over the last southern summer, Australia became the planetary symbol of a climate-changed future, the canary in the coal mine. In a nation suffering from unprecedented extremes of heat and aridity and defiantly expanding its coal-mining and export industry, bushfires swept across vast areas of coastal and inland forest, killing people, animals and trees. Bush and city-dwellers wore face masks as smoke engulfed their neighbourhoods for months, and rural folk trucked water and hay to struggling towns and farmers. The long summer of disaster lost any definition as an event and became, instead, a new and frightening normality.

Nature and culture were so intertwined that they could not be distinguished. Bushfire is integral to Australian ecologies and human histories, but what was this? As people suffered, survived and recovered, they looked for historical understanding of their experience: about what had just happened to them and why, about what they felt and remembered, and about how others had coped with such challenges in the past. This book gathers together scholarly wisdom on these questions and offers valuable research insights that are both practical and illuminating.

Histories of disaster are often propelled by compassion, a sense of urgency and a desire to be useful in times of grief and recovery. They understandably seek 'lessons from the past'. In this way they engage directly with one of the central issues of historiography: how does a scholar mediate responsibly between past and present, between pressing current needs and the integrity of past experience? The people who are the subjects of the scholarly studies in this volume are generally not distant and inscrutable; rather, they are living or passionately remembered and they shape or contest the histories written about them. Therefore, this is a book not only about the histories of disasters but also about disaster history, about the special challenges of writing historically about communities in crisis, even while the memories and debates are still volatile. It offers public history, oral history, environmental history and political history, seamlessly integrated as they must be and as experience testifies.

After Victoria's Black Saturday firestorm in 2009, I was a member of a team of historians invited by a small community to assist them in the disaster. We were the only emergency personnel in the district not wearing hi-vis jackets. We arrived after the first urgent wave of assistance and were struck by the community's statement that they didn't want any more material assistance, soft toys or hugs, but *did* want help in understanding and recording what they had experienced. They wanted to know what exactly happened on the day, what it all meant and how they would reinvent themselves. It was our job not to tell but to listen, and this made us unusual and useful. The chapters in this book do that essential civic and intellectual work, which remains undervalued by authorities.

Disasters are not always single recognisable events; they can be 'slow catastrophes', as historian Rebecca Jones calls droughts, their full significance sometimes identifiable only in retrospect. And they cast a long shadow, changing lives and communities forever, imprinting themselves on personal and collective memories and on patterns of commemoration. In this sense, argues Scott McKinnon, all disasters are slow. One of the impressive dimensions of this collection is the attention given to oral history and public storytelling, memorials and myth-making. Historians don't just seek out the testimony of witnesses, they also dredge the layers of interpretation and debate, oral and written, that accumulate over time. Disasters, it emerges, have after-lives that can be as disturbing as the events themselves.

Commissions of inquiry crank through their proceedings, media and governments spin their stories, and the blame game is endlessly played out; disasters distil and concentrate social debate and become political and cultural landmarks. As studies in this book show, the meaning of a disaster changes over time under the workings of memory, history and politics, and needs to be tracked through a continuing landscape of instability. A persistent insight from disaster histories is that society finds it hard to acknowledge the sheer enduring power of nature, and prefers political and technological solutions over deeper environmental or cultural adaptations. Arsonists, greenies or dam engineers are easier to blame than poor policy, environmental ignorance or cultural blindness. People also tend to separate nature and culture and are less ready to acknowledge the complex and sometimes fatal integration of society and environment, the embedded historical patterns of settlement that can create or amplify disaster. As Margaret Cook shows in her chapter, a strong reflex exists that humans should not be defeated by nature.

Disaster histories are rarely written by historians, and more often—as Ian Townsend observes in this volume—by disaster scientists and policy wonks. Thus, they tend to cherry-pick past facts for present purpose rather than submit themselves to the contextual, immersive analysis of the historian, a method that respects the complex integrity of the past. The flawed 'Stay or Go' fire policy that condemned so many people to death on Black Saturday was founded on bad history written by non-historians.[1] It was history that reflected the tendency of disaster experts and bureaucrats to look for historical evidence that supported their preferred management policy. And the Black Saturday Royal Commission, despite its 155 days of evidence, failed to recognise this endemic problem of research because it sought advice from the very same experts, the architects of the policy, rather than from independent historians. This is a dangerous cycle that is hard to break. This book offers a way out of it, provided the distinctive skills of the historian are given due respect. As Townsend argues here, 'Remembering a disaster accurately is literally a matter of life and death.'

A further reason to seek disaster histories from historians is their inclination to listen to the voice of experience, however surprising or uncomfortable it may be. History at its best is a subversive discipline because it

[1] For further analysis, see my "The Disturbing Logic of 'Stay or Go'", *Inside Story*, 22 November 2012, https://insidestory.org.au/the-disturbing-logic-of-stay-or-go/

releases narratives that run counter to the official or dominant interpretation; they are not always right, but sometimes they contain enduring truths. For example, people who live in the firestorm forests of Victoria have developed special words and phrases for the extreme fire behaviour they have experienced. But the same experts and professionals who developed the 'Stay or Go' policy disparaged the eyewitness accounts of people who survived the firestorms of 1851, 1926, 1939 and 1983. They convinced themselves that the unrehearsed narratives of experience were actually exaggerated fictions or 'myths' that needed to be dispelled by calm professional education and fire science. As recently as 2008, thoughtful fire officers—drawing on the science of grassfires—argued that there were no such phenomena as 'exploding houses' or 'firestorms' or 'fireballs' and that these were just the delirious inventions of victims. And they suggested that such untutored and emotive words falsely implied that 'bushfire is something beyond human control'. This professional disparagement of local testimony and tragic underestimation of the power of nature clearly reveal the psychological blinkers of history done by disaster experts.

That's why this book—rich as it is with the words of witnesses and the voice of local memory—is so important. It also showcases the best new scholarship in the field by historians, bringing together work on bushfire, floods, cyclones and earthquakes. The elements once lay outside the domain of human history, set aside as 'acts of God', but now the relationships between people and nature, society and the environment, are at the heart of humanist scholarship and of our politics, both local and planetary. The Anthropocene has engulfed us and nature and culture cannot be teased apart—and histories of disaster expose that amalgam for intelligent and urgent scrutiny.

Australian National University Tom Griffiths
Canberra, ACT
Australia

Contents

Introduction 1
Scott McKinnon and Margaret Cook

Part I Lessons from Past Disasters 19

"Best Forgotten": Black Saturday's Difficult Stories 21
Peg Fraser

Shaped by Fire: How Bushfires Forged Migrant Environmental Understandings and Memories of Place 41
Gretel Evans

Placing Memories of Unforgettable Fires: Official Commemoration and Community Recovery After the 2003 ACT Firestorm 59
Scott McKinnon

Part II Human Understandings of Disaster 77

"Shaken but Not Stirred": The Aftermath of Urban Disasters 79
Margaret Cook

Escaping Water: Living Against Floods in Townsville, North Queensland, from Settlement to 2019 99
Rohan Lloyd, Patrick White, and Claire Brennan

A Perfect Storm 119
Ian Townsend

Part III Legacies of the Past 137

Shallow Fire Literacy Hinders Robust Fire Policy: Black Saturday and Prescribed Burning Debates 139
Daniel May

Decolonising Settler Hazardscapes of the Waipā: Māori and Pākehā Remembering of Flooding in the Waikato 1900–1950 159
Meg Parsons and Karen Fisher

Where the Wild Things Were 179
Deb Anderson

Index 199

Notes on Contributors

Deb Anderson is a writer who spent more than a decade with the independent press in Australia and overseas, writing mainly for *The Age*, before joining Monash University as a lecturer in 2013. Her research explores the lived experience of extreme weather and understanding of climate change. She is the author of *Endurance: Australian Stories of Drought* (2014). http://profiles.arts.monash.edu.au/deb-anderson/

Claire Brennan is a history lecturer at James Cook University, Townsville, where she researches the environmental history of northern Australia. She is particularly interested in the history of hunting and the way in which animals are used to create connections between people and landscapes. At present she researches the history of crocodile hunting in northern Australia, although she is engaged in writing the history of northern Australia more broadly. Her research portfolio is available at https://research.jcu.edu.au/portfolio/claire.brennan/

Margaret Cook has been a freelance historian for many years. Her environmental history PhD and subsequent book, *A River with a City Problem: A History of Brisbane Floods* (UQ 2019), explored floods and the relationship between the Brisbane River and its human riverside dwellers. Her research interests include natural disasters, water, climate and the cotton industry. A history lecturer at University of the Sunshine Coast, she holds Honorary Research Fellow positions at La Trobe University and University of Queensland. http://margaretcookhistorian.com.au

Gretel Evans is a PhD candidate at the University of Melbourne and an Associate Student in the Bushfire and Natural Hazards Cooperative Research Centre. Her PhD thesis draws upon oral history and memory studies to examine migrants' memories of bushfires and floods in Australia. Gretel has participated in archaeology projects in Australia and Macedonia and has worked as a consulting historian. She was an interviewer and research assistant for the YCW oral history project and a 2016 Summer Scholar at the National Library of Australia.

Karen Fisher (Māori/Pākehā/Other, iwi affiliations: *Ngāti Maniapoto, Waikato-Tainui*) is an Associate Professor in the School Environment at the University of Auckland. She is a geographer with research interests in environmental governance and the politics of resource use. Her research involves applying feminist and postcolonial theories to explore the creation and implementation of new co-governance arrangements between Māori iwi (tribes) and the New Zealand government over the management of rivers in Aotearoa New Zealand. https://unidirectory.auckland.ac.nz/profile/k-fisher

Peg Fraser is a historian from Museums Victoria who in 2010 documented the Black Saturday bushfires in the small settlement of Strathewen, Victoria. She continued her work with bushfire survivors while completing her PhD at Monash University. She is the author of *Black Saturday: Not the End of the Story* (2018). Prior to the Bushfires Collection, Peg worked across many aspects of Museums Victoria's collections and exhibitions. She has written or curated exhibitions on Victoria's migrant communities, refugees and child migrants as well as material culture collections including nineteenth-century textiles, numismatics and scientific artwork. She holds an honorary research appointment with Museums Victoria.

Rohan Lloyd is a high school teacher and historian from Townsville who researches environmental history of North Queensland and the histories of environmentalism. Rohan's major research project, and the subject of his PhD from James Cook University, is the history of the Great Barrier Reef. He was a 2015 Summer Scholar at the National Library of Australia.

Daniel May is a PhD candidate in the School of History, Australian National University, and an Associate Student in the Bushfire and Natural Hazards Cooperative Research Centre. His PhD thesis investigates the historical and contemporary politics of Indigenous burning in Australia and the Western United States. In 2018 he completed an Endeavour Research Fellowship in the United States. Daniel's research is supported by an Australian Government Research Training Program (RTP) Scholarship. He regularly posts on fire-related matters at @DJMay19.

Scott McKinnon is a Vice Chancellor's Postdoctoral Research Fellow in the Australian Centre for Culture, Environment, Society and Space (ACCESS) at University of Wollongong. He is an oral historian and geographer with a research background in disasters research, geographies of memory and histories and geographies of sexuality and gender. Scott's postdoctoral project focusses on collective memories of disaster in Australia. He is the author of *Gay Men at the Movies: Cinema, Memory and the History of a Gay Male Community* (2016).

Meg Parsons (Māori/Pākehā/Lebanese, iwi affiliation: *Ngāpuhi*) is a senior lecturer in the School of Environment at the University of Auckland, Aotearoa New Zealand. She holds a PhD from the University of Sydney and has worked at the Waitangi Tribunal in New Zealand and the University of Melbourne and Griffith University in Australia. Her research focuses on understanding the drivers and long-term consequences of social and ecological transformations to Aotearoa New Zealand freshwater systems. She is a contributing author to the Sixth Assessment of Working Group II for the Intergovernmental Panel on Climate Change (IPCC). https://unidirectory.auckland.ac.nz/profile/meg-parsons

Ian Townsend holds a PhD in History from the University of Queensland with a focus on the 1899 Bathurst Bay cyclone. He worked for many years as an investigative journalist and broadcaster at ABC Radio National and is the winner of four Australian Museum Eureka

Prizes for science and medical journalism, as well as an Australian Human Rights Award. He has previously written two novels and a non-fiction book exploring three different Australian disasters.

Patrick White is a PhD candidate at James Cook University in Townsville. His research analyses how local governments in Queensland influenced post-war reconstruction and northern development in Australia. Patrick's work has been published in both academic journals and *The Conversation*. He has experienced a variety of natural disasters including Townsville's 'Night of Noah' floods in 1998 and severe tropical cyclone Yasi in 2011, and during the 2019 flooding of Townsville, his family was evacuated from their home after a landslide on a nearby mountain. Patrick also serves in the Australian Army and assisted on 'Operation Townsville Flood Assist' in 2019.

List of Figures

Introduction

Fig. 1 Map of Australian states 2
Fig. 2 Map of New Zealand 6

"Best Forgotten": Black Saturday's Difficult Stories

Fig. 1 Map indicating key locations in Victoria 22

Shaped by Fire: How Bushfires Forged Migrant Environmental Understandings and Memories of Place

Fig. 1 Map indicating key locations in Victoria 43

Placing Memories of Unforgettable Fires: Official Commemoration and Community Recovery After the 2003 ACT Firestorm

Fig. 1 Map indicating key locations in the ACT 63

"Shaken but Not Stirred": The Aftermath of Urban Disasters

Fig. 1 Map of locations mentioned in this chapter 81

Escaping Water: Living Against Floods in Townsville, North Queensland, from Settlement to 2019

Fig. 1 Map of key locations in Townsville and surrounds 101

A Perfect Storm

Fig. 1 Map of locations in the Cape York region, Far North Queensland 120

Shallow Fire Literacy Hinders Robust Fire Policy: Black Saturday and Prescribed Burning Debates

Fig. 1 Map of key locations in Victoria and Western Australia 140

Decolonising Settler Hazardscapes of the Waipā: Māori and Pākehā Remembering of Flooding in the Waikato 1900–1950

Fig. 1 Map of locations around the Waipā Valley region, Aotearoa (New Zealand) 160

Where the Wild Things Were

Fig. 1 Map of locations mentioned in Queensland 182

Introduction

Scott McKinnon and Margaret Cook

As we write this introduction in December 2019, large areas of eastern Australia are on fire. More than 70 bushfires are alight in New South Wales (NSW) and Queensland (see Fig. 1 for all locations mentioned). Hundreds of homes have been destroyed and several people have lost their lives. Social media is permeated with photographs of flaming landscapes and frightening red skies taken from backyards or apartment balconies. Sydney's world-famous Opera House and Sydney Harbour Bridge are barely visible through an orange, smoke-filled sky, as residents breathe toxic air. Satellite imagery shows plumes of smoke visible from space. Downloads of a "Fires Near Me" app developed by the NSW Fire Service are skyrocketing. As co-editors, each of us working on this book is aware

S. McKinnon (✉)
University of Wollongong, Wollongong, NSW, Australia
e-mail: scottmck@uow.edu.au

M. Cook
School of Social Sciences, University of the Sunshine Coast,
Sippy Downs, QLD, Australia
e-mail: mcook1@usc.edu.au

S. McKinnon, M. Cook (eds.), *Disasters in Australia and New Zealand*,
https://doi.org/10.1007/978-981-15-4382-1_1

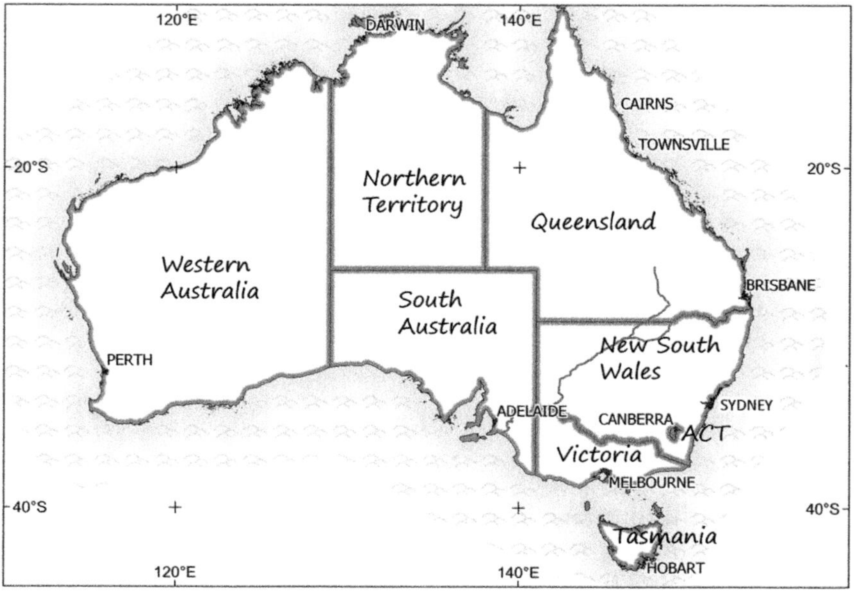

Fig. 1 Map of Australian states

that our own homes are not free from risk. Margaret Cook's neighbourhood is thick with smoke from fires in nearby bushland. The region in which Scott McKinnon lives, thousands of kilometres away, has been subject to a 'Catastrophic' level fire warning. Summer has only just begun.

It seems clear that a changing climate is creating conditions of increasing risk of disaster in Australia and around the world. It also seems at least possible that these fires will finally force the Australian government into taking real action to mitigate the effects of anthropogenic climate change. While the New Zealand government has recognised this issue as centrally important to the future of its nation and neighbours, its Australian counterparts remain reluctant to do so. Perhaps these widespread fires are, at last, a catastrophe that can't be ignored. And yet, as historians of disaster, we know that once these fires are extinguished and disappear from the front pages, they will quickly seem less urgent. For the people who have lost loved ones or homes, the recovery will likely be slow and challenging. But the rest of us will move on to new concerns, until the next time.

Those who deny the reality of climate change draw on history to bolster their claims. They argue that Australia has always experienced terrible bushfires, which is undeniably true; fire is, after all, a cyclical element of large parts of the Australian environment. But emergency management officials have described these fires as "unprecedented," urging the public to understand that increasing periods of hot weather means more days of higher fire risk, drier conditions in bushland and less time throughout the year in which to undertake mitigation strategies. Denialists also seek to blame 'green' environmental policies, claiming that the real problem is a lack of prescribed burning. Forceful opinions are expressed about burning as a fire management strategy, often by individuals with little apparent understanding of the history and complexity of this process.

In the words of sociologist Michelle Duffy and media scholar Susan Yell, "the Australian summer is framed by a narrative of bushfire."[1] Southern Australia is recognised as one of the world's worst bushfire-prone regions. With its forests of mountain ash, dependent on fire for regeneration, Tom Griffiths describes Victoria as "the most dangerous fire region on the planet."[2] While bushfires preoccupy the environmental imaginary in southern Australia's summer months, for northern NSW and Western Australia (WA), Queensland and Northern Territory (NT), those months (December to February) can also be flood season. Australia's climate is driven by the El Niño-Southern Oscillation (ENSO) cycles, which loosely operate over timescales between one and eight years. The warming Pacific Ocean in La Niña years increases flood risk.[3]

In northern Queensland, NT and WA, summer is cyclone season, the time when monsoonal troughs build offshore and head towards the coast. Most summers, an average of 13 tropical cyclones (known as hurricanes

---

[1] Michelle Duffy and Susan Yell, "Mediated Public Emotion: Collective Grief and Australian Natural Disasters," in *Emotions and Social Change: Historical and Sociological Perspectives*, ed. David Lemmings and Ann Brooks (New York: Routledge, 2016), 99.

[2] Tom Griffiths, "An Unnatural Disaster?" *History Australia* 6, no. 2 (2009), 35.2.

[3] Australian Government Bureau of Meteorology, "What Is La Niña and How Does It Impact Australia?" 2016, http://www.bom.gov.au/climate/updates/articles/a020.shtml.

in the northern hemisphere)[4] cross the coast of northern and western Australia, and half become severe.[5] While north western Australia is the county's most cyclone-prone region, Queensland is no stranger to devastating cyclones, with 4.7 on average per year and 208 known impacts since 1858.[6] Queensland's deadliest cyclone was *Mahina* that in 1899 devastated a pearling fleet north of Cooktown (see Townsend chapter). Only the two cyclones that occurred in 1918 rival Yasi in 2011 for second position in terms of intensity.[7]

The scale of events in Australia is hard to comprehend, although examples help bring home the enormity of some disasters. On 5 January 2011 more than 78 per cent of Queensland, over 1 million square kilometres (greater than the area of France and Germany combined) was flooded, leaving 40 towns and 2.5 million people affected.[8] By 9 November 2019 (before the start of the traditional bushfire season), over 850,000 hectares of NSW had been scorched by bushfire. One fire, west of Sydney, the Gospers Mountain fire, had by 20 December 2019 destroyed an area seven times the size of Singapore. Burning in excess 444,000 hectares this "mega fire" is the largest Australian forest fire to date from a single ignition point.[9]

---

[4] Cyclones are formed over the South Pacific and Indian Ocean. Hurricanes over the North Atlantic Ocean and Northeast Pacific and typhoons are over the Northwest Pacific Ocean. In the southern hemisphere they spin clockwise, anticlockwise in the northern hemisphere. All are rotating low-pressure weather systems of inwardly spiralling winds of over 104 km per hour.

[5] Bill Bunbury, *Cyclone Tracy: Picking up the Pieces* (South Freemantle, WA: Freemantle Arts Centre Press, 1994); Peter Reid, *Returning to Nothing: The Meaning of Lost Places* (Cambridge: Cambridge University Press, 2011); Sophie Cunningham, *Warning: Cyclone Tracy* (Melbourne, VIC: The Text Publishing Company, 2014); Patti Roberts, *Surviving Tracy: Cyclone Tracy Survivor Stories* (Place of publication not identified: Patti Roberts, 2015); and Julie Roberts and Martin Young, "Transience, Memory and Induced Amnesia: The Re-imagining of Darwin," *Journal of Australian Studies* 32, no. 1 (2008): 51–62.

[6] Australian Government, Bureau of Meteorology, "Tropical Cyclones in Queensland," 2019, http://www.bom.gov.au/cyclone/about/eastern.shtml.

[7] Australian Government, Bureau of Meteorology, "Severe Tropical Cyclone Yasi," 2019, http://www.bom.gov.au/cyclone/history/yasi.shtml.

[8] Catherine Holmes and Queensland Floods Commission of Inquiry (QFCI), *QFCI Final Report* (Brisbane: QFCI, 2012), 32.

[9] Harriet Alexander and Nick Moir, "'The Monster': A Short History of Australia's Biggest Forest Fire," *Sydney Morning Herald*, 20 December 2019, https://www.smh.com.au/national/nsw/the-monster-a-short-history-of-australia-s-biggest-forest-fire-20191218-p53l4y.html.

Although often overlooked, Australia does not escape earthquakes, commonly associated with the earthquake vulnerability zones in WA and South Australia (SA), but includes an area from Tasmania to northern NSW. In contrast Aotearoa New Zealand is known for earthquakes, with approximately 20,000 detected annually on seismographs (and between 100 and 150 felt), the country has earned its nickname the Shaky Isles (for New Zealand locations, see Fig. 2).[10] The 7.8 magnitude earthquake that destroyed Napier and Hastings in 1931 remains the nation's deadliest, as at least 256 people died, with thousands injured. Christchurch city is still rebuilding after two severe earthquakes in 2010 and 2011 that killed 185 people and left some areas unsafe to re-enter. Situated on the boundary of two tectonic plates, New Zealand is particularly vulnerable to earthquakes, tsunami, landslides and volcanic eruptions. The constant clouds of visible gases from White Island, the bubbling mud of Rotorua in the Bay of Plenty, and the rumblings of Mt Ruapehu in the middle of the North Island are reminders of the ever-present volcanic hazard.

On average ten tropical cyclones form in the South Pacific tropics every year, any one of which may affect New Zealand bringing severe storms and floods. Cyclone Bola in March 1988 remains the most costly, causing three deaths and mass evacuations of thousands of residents in the Hawke's Bay, Gisborne and East Cape region on the North Island. Subsequent floods in the Waipoa River destroyed bridges and property. As we write, floodwaters in Wanaka and Queenstown in New Zealand's South Island are threatening properties and isolating communities. The death toll from the eruption of White Island on 9 December has reached 16 with 2 people missing. New Zealand, it seems, is also heading for a summer of disasters.

In this volume, we seek a greater understanding of the impact of these disasters on Australia and New Zealand in the hope that by building knowledge of the past we might better understand the challenges of our uncertain present. Although some disasters are recalled as significant moments in the history of each country, their historiography remains

[10] GeoNet, Geological Hazard Information for New Zealand, https://www.geonet.org.nz/earthquake/statistics [accessed 17 September 2019].

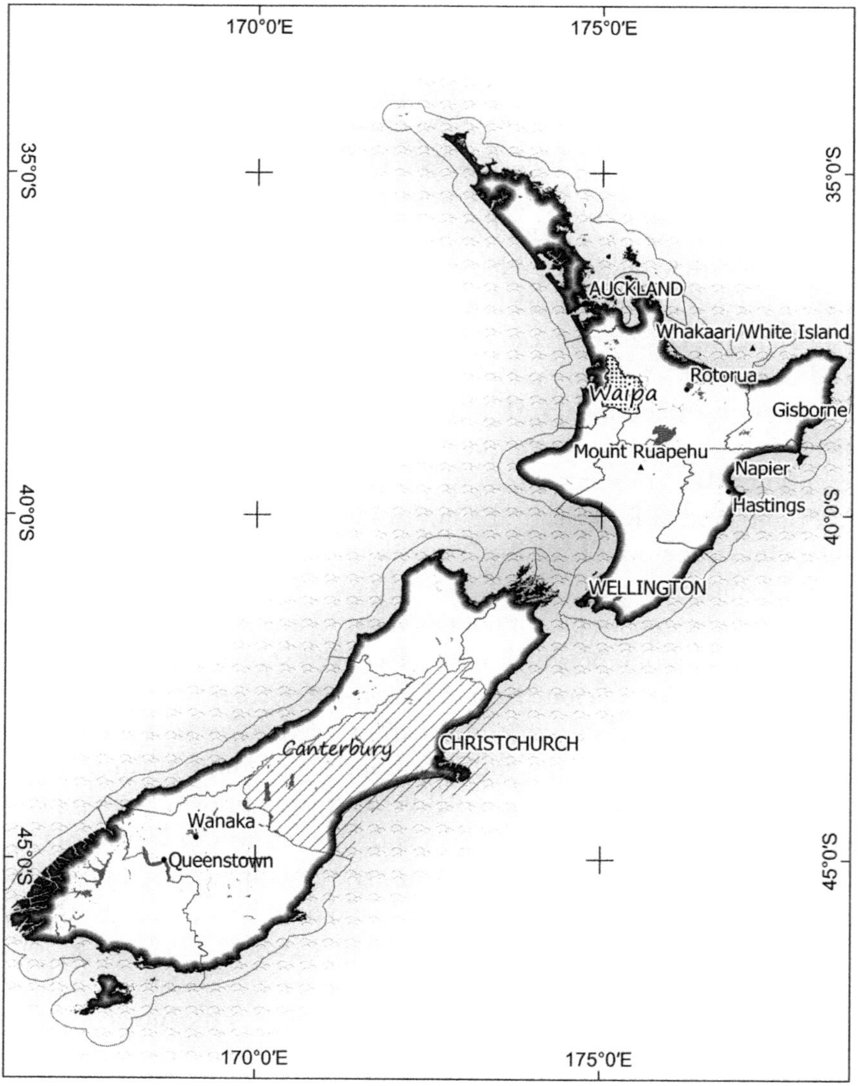

Fig. 2 Map of New Zealand

surprisingly limited. In 1999, Australian historian Paula Hamilton argued, "disasters are not usually regarded as the substance of scholarly work by historians; it is left rather to other storytellers to keep these tales alive in the cultural imagination."[11] This volume builds on and contributes to a growing body of recent scholarship addressing that significant historiographical gap, placing disasters within the cultural, social, political and environmental histories of each nation.

Defining Disasters

This collection contains histories of earthquake, flood, bushfire and cyclone, each of which would generally be described as a form of 'natural' disaster. As each of the chapters contained within reveals, however, disasters are not natural. While they may be triggered by a 'natural' event, the term disaster is anthropogenic. Without humans, a cyclone is just weather; an earthquake merely the movement of tectonic plates.

Environmental historian Tom Griffiths has described the 1939 Black Friday bushfires as "a cultural exaggeration of a natural rhythm."[12] In many landscapes, fire or flood is part of the environment's life cycle, potentially bringing rejuvenation. Research across a number of disciplines in the humanities and social sciences has identified the importance of reframing broader understandings of disaster to make this point clear: to call a disaster 'natural' is to risk masking or obscuring the role of human decisions in their production. A region's vulnerability to disaster is "partly the result of historical development paths and governance processes," determined by prevailing socio-economic values.[13]

---

[11] Paula Hamilton, "Memory Remains: Ferry Disaster, Sydney 1938," *History Workshop Journal* 47, no. 1 (1999), 197.

[12] Tom Griffiths, "How Many Trees Make a Forest? Cultural Debates About Vegetation Change in Australia," *Australian Journal of Botany* 50, no. 4 (2002), 387.

[13] Iain White, *Water and the City: Risk, Resilience and Planning for a Sustainable Future* (London: Routledge, 2010). See also Rohan D'Souza, *Drowned and Damned: Colonial Capitalism and Flood Control in Eastern India* (Oxford: Oxford University Press, 2006); Matthew Everard, *The Hydro-Politics of Dams* (London: Zed Books, 2013); Mark Pelling, "The Political Ecology of Flood Hazard in Urban Guyana," *Geoforum* 30, no. 3: 249–261; Ben Wisner, Piers Blaikie, Terry Cannon, and Ian Davis, *At Risk: Natural Hazards, People's Vulnerability and Disaster* (London: Routledge, 2003);

We have not included histories of drought in this volume,[14] but instead have focussed on comparatively rapid onset disasters, those that are generally assigned a fixed date. In the case of fire and cyclone they are also ascribed a name (Black Saturday, Ash Wednesday, Cyclone Tracy, Cyclone Yasi), locking them into a time and defining them as a cultural entity. They are regarded as climatic or environmental 'events.' And yet the time periods used to define these events by governments or for the purposes of emergency management are often more specific or fixed than we might use as historians. A flood, for example, might be defined as beginning with the arrival of excess water into a specific place, but may also require preceding wet periods (often months long), when the land is soaked and the runoff is higher. It also requires the decision of humans to occupy that place in ways that are incompatible with the intermittent flow of large amounts of water. The end of the flood may be declared when the waters recede, and yet the history of the disaster might also encompass the long recovery periods of survivors, or the ensuing political debates about flood planning, urban development or water management.

We argue, therefore, that the temporal boundaries of a disaster are blurry at best. A disaster should not be considered an immediate, quick event; a temporary aberration in a linear path of progress, that is soon forgotten. Disasters are both unpredictable and cyclical; inevitable and

Ben Wisner, "The Political Economy of Hazards: More Limits to Growth?" *Environmental Hazards* 2, no. 2 (2000): 59–61; Anthony Oliver Smith, "Theorizing Vulnerability in a Globalised World: A Political Ecology Perspective," in *Mapping Vulnerability: Disasters, Development and People*, ed. Greg Bankoff, Georg Frerks, and Dorothea Hilhorst (London: Earthscan, 2004), 10; Ruth Morgan, "Histories for an Uncertain Future: Environmental History and Climate Change," *Australian Historical Studies* 44, no. 3 (2016), 354.

[14] In recent decades drought has gained scholarly attention, particularly in the Mallee region of Australia which traverses parts of New South Wales (NSW), Victoria and South Australia (SA), and extends across the Nullarbor into Western Australia (WA). Notable examples include: Rebecca Jones, *Slow Catastrophes: Living with Drought in Australia* (Clayton, VIC: Monash University Publishing, 2017); Jenny Keating, *The Drought Walked Through: A History of Water Shortage in Victoria* (Melbourne: Department of Water Resources Victoria, 1992); Tom Griffiths, "Mallee Roots: A Brief History of Victoria's Northwest," *Park Watch* 178 (1994); Deb Anderson, *Endurance: Australian Stories of Drought* (Collingwood: CSIRO Publishing, 2014); Katie Holmes, "'It's the Devil You Know': Environmental Stories from the Victorian Mallee," in *Telling Environmental Histories: Intersections of Memory, Narrative and Environment*, ed. Katie Holmes and Heather Goodall (Cham, Switzerland: Palgrave Macmillan, 2017); Richard Broome, Charles Fahey, Andrea Gaynor and Katie Holmes, *Mallee Country: Land, People and History* (Melbourne: Monash University Publishing, 2019).

avoidable. They happen in moments, but their origins are hard to identify and their impacts endure over vast timescales. For survivors, disasters provide a biographical temporal marker, life before and after the disaster; they are at once both a singular *event* and a *process* of recovery with an indeterminate end. For the broader society, particular disasters may be adopted into local, regional or national identities, in which recovery from the disaster's impacts is celebrated as indicative of positive traits across the community. Equally, they may be largely forgotten, except by those with direct experience.

The spatial boundaries of disaster are equally complex. If an earthquake in Christchurch impacts the lives of individuals, transforms the city, and shapes the national economy of New Zealand, at what scale is the geography and the history of the disaster best understood? And how do we encompass the uneven and inequitable impacts of urban transformation or economic change across various communities within that city or nation? Geographer Kenneth Hewitt has argued against understanding disasters "as if these 'events' and all that happens within each one belong not only to a separate domain but to a single, albeit 'multivariate' reality."[15] Disasters don't happen *to* a society but originate within it, and their effects are determined as much by the structures of that society as they are by water, wind or flame.

Historiography

Until the 1970s disasters were commonly overlooked by historians in Australia and New Zealand, left to the remit of the sciences (seismologists, climatologists and geologists), the social sciences (geography and sociology) or disaster management. It was, in part, this disciplinary division in the USA that Hewitt was responding to when he argued in 1983 that "natural disaster is quarantined in thought as well as practice" and had become an "archipelago of isolated misfortunes."[16] Disasters were

---

[15] Kenneth Hewitt, *Interpretations of Calamity: From the Viewpoint of Human Ecology* (Winchester, MS: Allen & Unwin), 14.

[16] Hewitt, *Interpretations of Calamity*, 12.

considered random and singular events, rather than reflecting wider historical patterns, until Stephen Pyne and Tom Griffiths reversed this trend in Australia with their environmental histories of bushfires.[17] Fire scholarship paved the way to writing histories of disasters and primarily dealt with Australia's southern states, notably the regions of Victoria, the Australian Capital Territory (ACT) and Tasmania.[18] Perhaps the severity of the events and loss of life in these states explains the comparatively high scholarly attention given to bushfires, in comparison to other disasters.

In New Zealand, Catherine Knight's environmental history of rivers, although not specifically a history of disaster, has nonetheless addressed flooding and flood control. The Napier and Christchurch/Canterbury earthquakes have received scholarly attention but generally through the prism of hazard vulnerability and resilience.[19] The work of Lisa Langer and Tara McGee offers an inter-generational Māori cultural perspective on fire, now complemented by Meg Parsons and Karen Fisher's research on Māori understandings of flood.[20] However, New Zealand disasters are a field ripe for further work.

In Australia, Emily O'Gorman's *Flood Country* provides the first detailed catchment study of floods covering 150 years, using four key floods in the Murray-Darling system to "illuminate changing ways of

---

[17] Stephen J. Pyne, *Burning Bush: A Fire History of Australia* (New York: Henry Holt and Company, 1991); Tom Griffiths, *Forests of Ash: An Environmental History* (Cambridge: Cambridge University Press, 2002).

[18] See Christine Hansen and Tom Griffiths, *Living with Fire: People, Nature and History in Steels Creek* (Melbourne: CSIRO Publishing, 2012); Peg Fraser, *Black Saturday: Not the End of the Story* (Melbourne: Monash University Press, 2018); Grace Karskens, "Fire in the Forests? Exploring the Human-Ecological History of Australia's First Frontier," *Environment and History* 25, no. 3 (2019): 391–419.

[19] Eric Pawson, "Environmental Hazards and Natural Disasters," *New Zealand Geographer* 67, no. 3 (2011): 143–147; Katie Pickles, *Christchurch Ruptures* (Wellington: Bridget William Brooks, 2016). For examples of risk and vulnerability research, see J. C. Gaillard and Christopher Gomes, "Post-disaster Research: Is There Gold Worth the Rush?" *Jàmbá: Journal of Disaster Risk Studies* 7, no. 1 (2015), https://doi.org/10.4102/jamba.v7i1.120; Mischa Hill and J. C. Gaillard, "Integrating Disaster Risk Reduction into Post-disaster Reconstruction: A Long-term Perspective of the 1931 Earthquake in Napier, New Zealand," *New Zealand Geographer* 69, no. 2 (2013): 108–119.

[20] E. R. Langer and Tara K McGee, "Wildfire Risk Awareness and Prevention by Predominantly Maōri Rural Residents, Karikari Peninsula, Aotearoa New Zealand," *International Journal of Wildland Fire* 26 (2017): 820–828.

knowing and managing rivers, floods and floodplains."[21] Tim Bonyhady has examined colonial settler relationships with the Darling River, as has Grace Karskens in the Hawkesbury-Nepean region.[22] Jessica Weir and Heather Goodall highlighted differences in Indigenous and settler understandings of rivers in the Murray-Darling Basin, while Chas Keys studied Maitland floods and NSW flood management.[23] Margaret Cook has published on floods in the sub-tropical catchment of the Brisbane River, but other than Rockhampton, floods have been largely overlooked in the historiography of the rest of Queensland, a state frequently affected by floods.[24] Rohan Lloyd, Patrick White and Claire Brennan's chapter on Townsville floods offers a rare contribution to the rest of the state's flood historiography.

Cyclones, other than Cyclone Tracy that devastated Darwin in 1974, have been largely overlooked by Australian historians although overseas they have been given greater attention. Australian earthquakes have also been largely ignored, but perhaps not surprisingly given their severity, the Christchurch earthquakes of 2010 and 2011 have gained significant scholarly attention in New Zealand, especially in the work of historian Katie Pickles.[25]

The gradual positioning of disaster as an important issue for historical enquiry has largely occurred within the subfield of environmental history. This is understandable given that disasters sit within the intertwining of environment and culture that environmental historians see as the

---

[21] Emily O'Gorman, *Flood Country: An Environmental History of the Murray-Darling Basin* (Collingwood, VIC: CSIRO, 2012).

[22] Tim Bonyhady, *The Colonial Earth* (Carlton, VIC: Melbourne University Press, 2002); Grace Karskens, "Floods and Flood-Mindedness in Early Colonial Australia," *Environmental History* 21, no. 2: 315–342.

[23] Jessica Weir, *Murray River Country: An Ecological Dialogue with Traditional Owners* (Canberra: Aboriginal Studies Press, 2009); Heather Goodall, "The River Runs Backwards," in *Words for Country: Landscape and Language in Australia*, ed. Tim Bonyhady and Tom Griffiths (Sydney: UNSW Press, 2002); Chas Keys, *Maitland, City on the Hunter: Fighting Floods or Living with Them?* (Paterson, NSW: Hunter-Central Rivers Catchment Management Authority, 2008).

[24] Margaret Cook, *A River with a City Problem: A History of Brisbane River Floods* (Brisbane: University of Queensland Press, 2019); Barbara Webster, *Marooned: Rockhampton's Great Flood of 1918* (Brisbane: Cooperative Research Centre for Coastal Zone, Estuary and Waterway Management, 2003).

[25] Pickles, *Christchurch Ruptures*.

core of their research. The impacts of climate change have added increased urgency to environmental histories, and the broader environmental humanities, exploring the lessons we might draw from past disasters in confronting new challenges. In the words of Ruth Morgan, "The climatic conditions that we will face in the future will be new to us, but change itself is not new and environmental histories can shed light on how humans have responded to and understood environmental change in the past."[26] Environmental histories have the capacity to reveal the anthropogenic origins of disaster without creating false distinctions between the human and natural worlds.

We wonder, however, what might also be valuably contributed to this subject by historians with other specialities. In recent years, the work of oral historians and memory scholars, for example, has begun to offer insightful pathways of investigation, particularly in terms of how survivors place disaster within life narratives, and the differing experiences of disaster across various groups within communities. In *Telling Environmental Histories*, Katie Holmes and Heather Goodall note that oral history aids environmental history by examining the "intimate relationships between people and place, between individual and cultural memory, and their webs of connection with the natural world."[27] In this volume, we further highlight the value of oral history with work by researchers—Anderson, Evans, Fraser and McKinnon—who do not define their work as 'environmental history' but who nonetheless reveal individual accounts that complicate, challenge or disprove dominant rhetoric, offering a nuanced understanding of human and non-human inter-relationships.

Research into disaster memory also troubles the temporal boundaries of disaster, exploring how lost objects or sites of memory are interwoven with the trauma of disaster; examining how and why disasters take their place within collective memory; as well as the complexities of maintaining memory of past disasters in communities prone to bushfire, cyclone or flood. Peter Read's *Returning to Nothing*, for example, sensitively and evocatively explores how the loss of place is an enduring impact of

[26] Morgan, "Histories for an Uncertain Future," 34.

[27] Holmes and Goodall, *Telling Environmental Histories*, 2.

disaster.[28] Tom Griffiths has noted the "dangerous mismatch between the cyclic nature of fire and the short-term memory of communities"[29] in regional Victoria, while Scott McKinnon has traced how cycles of flooding often sit uncomfortably in the collective memory of Brisbane, Queensland.[30] Margaret Cook has argued that myths of flood immunity, designed to uphold the property market and the political elite in Southeast Queensland, have eroded both the residents' memory and resilience.[31] All remind us that collective memory is a social construct, which shapes and interprets the past.[32]

Whose Disaster?

An increasingly acknowledged truth within disasters research, and one which we believe offers a rich field of exploration by historians across the discipline, is the fact that disasters impact on different groups within communities in varying ways. In other words, social marginalisation and economic inequity as experienced in 'everyday' life will also define the experience of disaster for many.[33] Recent work in the humanities and social sciences—including by contributors to this volume—has begun to explore new pathways that ensure greater inclusivity in understandings of the social impacts of disaster, thereby adding greater complexity to histories of disaster.

Both Australia and New Zealand are settler-colonial nations and a critical space of enquiry is exploration of Indigenous experiences of

---

[28] Reid, *Returning to Nothing.*

[29] Griffiths, "An Unnatural Disaster," 35.5.

[30] Scott McKinnon, "Remembering and Forgetting 1974: The 2011 Brisbane Floods and Memories of an Earlier Disaster," *Geographical Research* 57, no. 2 (2019): 204–214.

[31] Margaret Cook, "'It will never happen again': The Myth of Flood Immunity in Brisbane," *Journal of Australian Studies* 42, no. 3 (2018): 328–342.

[32] Wulf Kansteiner, "Finding Meaning in Memory: A Methodological Critique of Collective Memory Studies," *History and Theory* 41, no. 2 (2002): 179–197; Wendy Madsen and Cathy O'Mullan, "Responding to Disaster: Applying the Lens of Social Memory," *Australian Journal of Communication* 40, no. 1 (2013): 57–90.

[33] Ben Wisner, "Marginality and Vulnerability: Why the Homeless of Tokyo Don't 'Count' in Disaster Preparations," *Applied Geography* 18, no. 1 (1998): 25–33.

disaster, both pre- and post-colonisation. Work in this field has examined differing epistemologies of fire and flood among Indigenous inhabitants and colonisers. In their chapter Meg Parsons and Karen Fisher argue that for the Māori people of the Waipā River region, floods were considered times of vital nourishment for the land. For Māori, colonisation was the disaster, not the flood. Work in the social sciences has also explored Māori experiences of more recent disasters, particularly the 2011 Christchurch earthquakes.[34] This work reveals both the added layers of vulnerability experienced by a marginalised community, as well as the resilience of that community developed through community support structures and traditional knowledges.

In Australia, Indigenous burning practices have been the subject of increasing study, both for understanding the history of Aboriginal people's relationship to country and as a source of vitally important strategies for fire management. In this volume, Daniel May explores how these practices became part of angry debates about prescribed burning policies in Victoria following the 2009 Black Saturday fires. Geographers and other disasters researchers have explored the increasing use of Indigenous fire knowledge in present-day fire mitigation strategies.[35] Yet Indigenous people's experiences of more recent disasters remain under-investigated to this point in Australia. One exception to this is recent research into the impacts of Cyclone Tracy on Aboriginal communities in and around Darwin in 1974.[36] Ian Townsend's chapter addresses the challenges of understanding historical disaster impacts on communities whose experiences were not considered worthy of recording at the time of the disaster.

---

[34] See, for example, S. Phibbs, C. Kenney, and M. Solomon, "Ngā Mōwaho: An Analysis of Māori Responses to the Christchurch Earthquakes," *Kōtuitui: New Zealand Journal of Social Sciences Online* 10, no. 2 (2015): 72–82; Simon J. Lambert, "Indigenous Peoples and Urban Disaster: Maori Responses to the 2010–12 Christchurch Earthquakes," *Australasian Journal of Disaster and Trauma Studies* 18, no. 1 (2014): 39.

[35] Christine Eriksen and Don L. Hankins. "The Retention, Revival, and Subjugation of Indigenous Fire Knowledge Through Agency Fire Fighting in Eastern Australia and California," *Society & Natural Resources* 27, no. 12 (2014): 1288–1303.

[36] Katherine Haynes, Deanne K. Bird, Dean Carson, Stephen Larkin, and Matthew S. Mason, "Institutional Response and Indigenous Experiences of Cyclone Tracy," National Climate Change Adaptation Research Facility, 2011.

Research in the social sciences also points us towards the need for greater examination of gender, class and sexuality within disaster histories. Popular histories of disaster often highlight narratives of 'communities pulling together' and successfully overcoming nature's attacks through cheerful and determined common purpose. While extraordinary acts of bravery and unity deserve their place in these histories, such narratives equally risk masking the impacts of misogyny and other forms of discrimination on the lives of many survivors. We know, for example, that rates of domestic violence often increase in communities struck by disaster.[37] Women also often face bullying and harassment when working in emergency management.[38] Equally, recent research has identified how homophobia, biphobia and transphobia impact on the lives of LGBTQ disaster survivors, including accessing vital disaster support services.[39] Postcolonial, feminist and queer historians have begun to place issues of race, gender and sexuality within disaster histories; however, there remains significant potential to continue to build on this work.

Histories of Disaster in Australia and New Zealand

This volume emerged from a workshop co-convened by the editors and held at the University of Wollongong, on Dharawal Country, in July 2019. We first met through a common interest in the history of flooding in Brisbane and Southeast Queensland and mutual surprise at the limited historiography of disaster in Australia and New Zealand. We felt that it was a timely moment to bring together scholars working in this field to discuss their research, to develop collegial connections, and to contemplate future paths for writing disaster histories. Over two days, nine

[37] Debra Parkinson and Claire Zara, "The Hidden Disaster: Domestic Violence in the Aftermath of Natural Disaster," *Australian Journal of Emergency Management* 28, no. 2 (2013): 28–35.

[38] Christine Eriksen, "Negotiating Adversity with Humour: A Case Study of Wildland Firefighter Women," *Political Geography* 68 (2019): 139–145.

[39] Dale Dominey-Howes, Andrew Gorman-Murray, and Scott McKinnon, "Queering Disasters: On the Need to Account for LGBTI Experiences in Natural Disaster Contexts," *Gender, Place & Culture* 21, no. 7 (2014): 905–918.

researchers met, sitting in the shadow of a tree-filled escarpment with its own history of fire and on a campus with a history of flooding. Participants came from New Zealand and the Australian states of Victoria, NSW and Queensland and the ACT. The group included environmental historians, historical geographers, oral historians and historians of migration, offering richly diverse approaches to studies of bushfire, flood, cyclone and earthquake. The workshop critically reviewed and reworked each of the papers that provide the majority of the work within this collection.

The chapters move through time and space, from the 1899 Cyclone *Mahina* to 2019's Townsville flood and from the Waipā River catchment of Aotearoa New Zealand's North Island to far northern Australia. The authors have drawn on varied methodologies—interviews, archival research, field work and text and object analysis—as well as their own scholarly training in museum studies, journalism, oral history, environmental history or human geography to unravel specific disasters and tell their stories.

We have divided the book into three sections. Part I, "Lessons from Past Disasters," has three chapters which draw on oral histories, as well as other sources, to contemplate the ongoing place of bushfires in the lives of survivors. Peg Fraser confronts the substantial challenges of working with a community devastated by fire and considers the role of the oral historian in revealing—or not—difficult stories of community fracture. Gretel Evans highlights the narratives of migrants to Australia and the ways in which fire both disrupted and developed their understandings of the Australian environment. Scott McKinnon traces the first three years after a bushfire in the ACT, focussing on anniversary events and sites to understand how the local community has placed memories of the fire.

In Part II, "Human Understandings of Disaster," Margaret Cook compares the history of two urban disasters, exploring how cities are often defined by a sense of control over nature, an idea that is shaken by earthquakes and floods. Rohan Lloyd, Patrick White and Claire Brennan offer a concise history of flooding in Townsville since colonisation, revealing a constant interplay between flood and water scarcity in the city's history. Ian Townsend also investigates disaster histories in Australia's tropical north, offering a valuable challenge to scientific data on Australia's deadliest cyclone through in-depth archival research.

In Part III, "Legacies of the Past," Daniel May examines the history of prescribed burning as a fire mitigation strategy in Victoria, providing a nuanced analysis of an angry debate. Meg Parsons and Karen Fisher explore the history of flooding in the Waipā River catchment area, revealing conflicts between Māori and Pākehā understandings of the environment, and the ongoing legacies of settler colonial projects. Deb Anderson investigates local knowledges and identities in communities in North Queensland facing increased disaster risk as a result of climate change. Importantly her chapter shifts the anthropogenic gaze of this book to the non-human world of the glider and the Rufus owl. While humans struggled to adjust to their altered sense of place, for the gliders the loss of habitat from the disaster threatened the survival of its species.

Each chapter makes a timely intervention into the public discourse on disasters, disrupting accepted narratives, challenging myths and memories and reminding us of the escalating hazards created through human actions. Today we face the unprecedented challenges of a global environmental crisis, a catastrophe well beyond the scope of an individual bushfire, flood or cyclone. There is an urgency for scholarship that helps the public understand the enormous implications of climate change and its immediacy.[40] It is hoped that these accounts of the interaction of the human and non-human world through the lens of disasters will increase awareness of both the seriousness and complexities of the environmental challenges that face us now, so that we may mitigate the risk in the future.

[40] Tom Griffiths, "The Planet Is Alive: Radical Histories for Uncanny Times," *Griffith Review* 63 (2018): 67.

Part I

Lessons from Past Disasters

"Best Forgotten": Black Saturday's Difficult Stories

Peg Fraser

It is, perhaps, a little absurd to begin a history with the words "Best forgotten". This is especially true of disaster history, where there is an additional impetus to remembering in the hope that we can learn something from the past to help predict, prevent or mitigate future fatal events.

Black Saturday, 9 February 2009, was one such disaster: a catastrophic bushfire that swept through the state of Victoria, Australia, killing 173 people (Fig. 1). In a series of important essays published in the years after Black Saturday, Tom Griffiths[1] evoked the memory of Judge Leonard Stretton, whose report on fatal bushfires in 1939 is the foundation for bushfire studies in Australia. In what has become the most quoted line from the report—"They had not lived long enough"—Stretton summed

[1] Tom Griffiths, "We Have Still Not Lived Long Enough," *Inside Story*, 16 February 2009, https://insidestory.org.au/we-have-still-not-lived-long-enough/

P. Fraser (✉)
Museums Victoria, Melbourne, VIC, Australia

S. McKinnon, M. Cook (eds.), *Disasters in Australia and New Zealand*,
https://doi.org/10.1007/978-981-15-4382-1_2

Fig. 1 Map indicating key locations in Victoria

up an essential problem of European occupation of the Australian continent: our failure to learn and remember.[2]

Griffiths argued that Stretton's remark was still relevant, referring to not only the disjuncture between the human life span and the cycle of catastrophic fires but also to the failure of a young, impatient settler colonial culture with only 200 years of living on this continent to understand the rhythm and rules of life here. Millennia of Indigenous land management through both agriculture and controlled burning were ignored, dismissed or destroyed.[3] Griffiths tracked the "language of forgetting", with its use of terms such as "unprecedented" and "unnatural", and the danger that future fires will also take people by surprise.[4]

[2] Judge Leonard Stretton, "1939 Bushfires Royal Commission: Report and Transcript of Evidence," University of Melbourne Archives, 5.

[3] For a history of Indigenous land management in Australia, see Bill Gammage, *The Biggest Estate on Earth: How Aborigines Made Australia* (Sydney: Allen and Unwin, 2012); Bruce Pascoe, *Dark Emu: Aboriginal Australia and the Birth of Agriculture* (Broome: Magabala Books, 2018).

[4] Tom Griffiths, "The Language of Catastrophe: Forgetting, Blaming and Bursting into Colour," *Griffith Review* 35, 2012.

I agree with the need to record experiences through disasters such as catastrophic bushfire. As a historian I am committed to the documentation and analysis of difficult material—I have worked with minority groups, with refugees, with child migrants and most of all with bushfire survivors[5]—but as an oral historian who works with intimate personal stories, frequently of a distressing nature, I worry about the complexities and issues of documenting challenging material.

This chapter is a discussion of the craft of oral history in times of disaster, a personal reflection on some of the ethical and methodological questions encountered while working with the Victorian Bushfires Collection at Museums Victoria. What stories do we share with the public and what do we leave untold? How do we balance competing interests in public education, community rebuilding and personal recovery? How do we deal with counter-histories, unintended revelations or conflicting accounts? And are there some stories that are just too difficult to tell? Perhaps it is true that, in the words of one narrator, "some things are best forgotten".

Black Saturday

It is now just over ten years since the fires of Black Saturday. Bushfires began in mid-January 2009, after ten years of drought and in the middle of the hottest summer then on record. Saturday 7 February was the most destructive day as uncontrolled fire swept through vast tracts of bushland, obliterated small towns and threatened the outer suburbs of Melbourne. The scale of the disaster unfolded over the following days: 173 people killed; more than 2000 homes destroyed; 78 communities affected; 430,000 hectares of land burnt; more than 8000 livestock and domestic

---

[5] See exhibition catalogues: Museums Victoria, "Stolen Childhoods," (Melbourne: Museums Victoria, 2011), https://museumsvictoria.com.au/article/stolen-childhoods/; Museums Victoria, "Talanoa: Stories of the Fiji Community," (Melbourne: Museums Victoria, 2009); Museums Victoria, "Handing on the Key: Palestinians in Australia," (Melbourne: Museums Victoria, 2009); Museums Victoria, "Survival of a Culture: Kurds in Australia," (Melbourne: Museums Victoria, 2008); Museums Victoria, "Trailblazers: Migrant women activists," (Melbourne, Museums Victoria, 2008); Museums Victoria, "Honouring Our Ancestors: Remembering East Timor," (Melbourne, Museums Victoria, 2007).

animals lost and untold numbers of native wildlife killed.[6] These were the most fatal fires in Australia's history.

Black Saturday was the latest in a long line of catastrophic bushfires documented since the early years of the colony of Victoria. Bushfire—an Australian word for a phenomenon known elsewhere in the world as wildfire—is one of the defining features of life in southeastern Australia. Stephen Pyne, an international authority on wildfire, calls Australia "the fire continent" and says that the southeastern part of the country is one of three places on earth where fire has shaped and continues to dominate the relationship between people and the environment.[7] With the effects of climate change, longer and more severe droughts and our changing settlement patterns, bushfires will only become more frequent and more destructive.

The Victorian Bushfires Collection

Soon after Black Saturday, Museums Victoria established the Victorian Bushfires Collection to preserve objects and oral histories connected to the State's history of bushfire. The museum is the state collection for Victoria and includes Indigenous Studies, natural history, social history and technology. The new bushfires collection fit in well with many existing collecting interests, including rural life and sustainable futures. It also follows an international trend to begin oral history projects in the wake of disaster, both man-made and natural.[8]

---

[6] Bernard Teague, Ronald McLeod and Susan Pascoe, *2009 Victorian Bushfires Royal Commission Final Report* (Melbourne: Parliament of Victoria, 2010), XXIII.

[7] Stephen Pyne, *Burning Bush: A Fire History of Australia*, 2nd ed., (Sydney: Allen & Unwin, 1992).

[8] See, for example, Mark Cave and Stephen Sloan, eds., *Listening on the Edge: Oral History in the Aftermath of Crisis* (New York: Oxford University Press, 2014); Mary Marshall Clark, "Herodotus Reconsidered: An Oral History of September 11, 2001, in New York City," *Radical History Review*, no. 111, Fall (2011): 79–89; Erin Jessee, "The Limits of Oral History: Ethics and Methodology Amid Highly Politicized Research Settings," *Oral History Review* 38, no. 2 (2011): 287–307; Chrischené Julius, "'Digging Deeper Than the Eye Approves': Oral Histories and Their Use in the "Digging Deeper" Exhibition of the District Six Museum," *Kronos*, no. 34(2008): 106–138; Carolyn Lunsford Mears, "A Columbine Study: Giving Voice, Hearing Meaning," *Oral History Review* 35, no. 2 (2008): 159–175.

In 2010 the museum started a targeted project with the aim of expanding its collection of bushfire-related objects, documents, images and oral histories. They assigned two curators, Rebecca Carland and me, to work with Liza Dale-Hallett, the senior curator of Sustainable Futures and founding curator of the Bushfires Collection. Our goal was to document, as far as possible within the limited time frame, the many faces of this event: the fire itself; the emergency response; the aftermath; the recovery efforts, and the ongoing challenges.

Liza continued her state-wide sampling of bushfire experiences, from the western districts to the northeast of the state. She covered as broad a geography and range of experience as possible: farmers and tree-changers; long-time residents and newcomers; emergency personnel and community leaders; people who had survived the fires and the relatives of those who had not. She also collected an extraordinary array of bushfire-related objects, from a seven-metre brick chimney to a pile of ash that had once been someone's tax records.

Bec focussed on linking Black Saturday to the museum's existing collections, mining the rich resources of artefacts, specimens and documents to develop connections that enhance our understanding of bushfire as a force that has long shaped our natural and social history, from deep time up to the present day.

While Liza and Bec were looking at the big picture of bushfire, I was interested in microhistory. I wanted to go deeply into one small place, to collect different stories and different perspectives of the same event, and to try to understand the variety and complexity of people's response to bushfire. That desire led me to Strathewen, a settlement of just 200 people on the northeast outskirts of Melbourne, that was hard-hit on Black Saturday. In one afternoon, more than 10 per cent of its population died and 80 per cent of its buildings were destroyed. In a remarkable display of generosity, 25 people—residents, former residents, bereaved relatives and relief workers—agreed to in-depth interviews, sometimes a series of interviews, that addressed not only the day of the fire but a wide range of personal experiences and attitudes. This was at a time when many of them were still attempting to rebuild, from almost nothing, the entire structure of their lives and I continue to be grateful to all the narrators for taking part. After our collecting project finished at the end of 2010, I continued working with Strathewen survivors as a Museums Victoria research associate and a PhD candidate at Monash University. The results were published in 2018,[9] but the work is ongoing.

The Oral History Interviews

All the interviews offered valuable first-person accounts, insights into not just the physical experience of surviving a major bushfire but also the practical, emotional and social impacts of the event. They are all 'difficult' stories, in that they contain distressing material, but many of them also raised difficult questions for us as curators. Museums Victoria had oral history guidelines developed for existing collections, but none of those earlier projects had contained the same level of intensity or emotional impact.

The interviews were semi-structured, allowing narrators to determine what would be discussed, and included life before and after Black Saturday. In preliminary meetings I discussed with each person what was and was not 'on the table'; years after the fires, some people were still not ready to talk about that day. I also brought up the emotional effects of being interviewed and some strategies for coping with that, as well as the narrator's rights with regard to use of the material. In some cases, I had to advise on the legal risks of making certain statements in an interview that would become a public (if not publicly available) document.

It seemed like every day we encountered new problems that had to be solved and it quickly became clear how important it was to be working in a team. We shared the emotional and practical challenges of working with bushfire survivors; consultation, debriefing and black humour were essential. We were aware of the impact of the material on curators, collection managers and conservators. The museum's offer of psychological counselling for curators was extended to our transcriber and to other members of staff who listened to the material. Many were Victorians who had personal connections to Black Saturday or earlier bushfires.

[9] Peg Fraser, *Black Saturday: Not the End of the Story* (Melbourne: Monash Publishing, 2018).

Part of our mission was to make the stories of survivors available to the public through Collections Online,[10] so from the beginning there was a built-in mandate to publish narratives as well as objects. This meant our work would quickly become part of the public domain. We set up new protocols for the accession, storage, transcription and access of the sensitive and sometimes disturbing material. These protocols were expanded and updated as new, unanticipated challenges arose.

The bushfire curators collected a wide range of stories, across different narrators and experiences but also with the same narrators across time. Stories changed as intervening time and events altered the narrator's interpretation. One narrator noted that her reactions to Strathewen moved in the years after Black Saturday, from anger at the actions of other residents to acknowledging that everyone was finding their own path through their grief. I realised that the permanent records of the interviews in the museum's collection had to be understood as documents at a point in time and that people had the capacity and the right to grow and change. This fluidity can be initially frustrating to those of us working with the stories but recording those changes not only adds to our own knowledge but also acknowledges that disaster recovery is an uneven and ongoing process. The work of clinical psychologist Rob Gordon, who has been working with disaster survivors since the Ash Wednesday bushfires of 1983, was particularly valuable in understanding the trajectory of recovery.[11]

Developing new processes took time and care, but for me the most challenging decisions were those made about the public use of the material. All of the people I interviewed gave me permission to use any part of their interviews in publications, without reservation. They also chose to use their real names, which was a source of both pride and anxiety, sometimes in the same person. We consulted with narrators about what to do with the material but in the end it was a curatorial decision how to

---

[10] Museums Victoria, "Victorian Bushfires Collection," https://collections.museumvictoria.com.au/articles/3032.

[11] Rob Gordon, "Community Impact of Disaster and Community Recovery," *InPsych: The Bulletin of the Australian Psychological Society,* 31, no. 2 (June 2009): 12–13; Rob Gordon, "Thirty Years of Trauma Work: Clarifying and Broadening the Consequences of Trauma," *Psychotherapy in Australia* 13, no. 3 (May 2007): 12–19.

interpret it. We respected the overarching goals of the collection but also had to take into account individual narrators' personal circumstances, the possible impact of publication on the individual and their community and the intended benefit to the larger community in telling these stories.

By delving deeply into one small place in a time of loss and recovery, I discovered wonderful stories about resilience, kindness and courage, but also stories of old injuries, new rivalries and a lot of unresolved anger. I listened to many people about the same event, using interviews and questions not only about what they saw and did but also about what they thought and felt, not only about one day but about their lives and relationships before and since.

The Poetry Tree

On my first day at work on the Bushfires Collection I saw a photo of the Poetry Tree, a burned manna gum in the heart of Strathewen that was used as a spontaneous memorial. The tree was covered with poems about Strathewen, the destruction and the long road to recovery, as well as photos of and tributes to deceased residents. Over time it had become a place of commemoration and a symbol of resilience for many people.

The texts and images on the tree created a narrative about Strathewen before Black Saturday as a place of beauty, peace and harmony. They told of an idyllic past interrupted by catastrophe but emphasised continuity and commitment to Strathewen. Through the local recovery committee, the Strathewen Community Renewal Association, I was introduced to the people who started the tree—Rosemary McKimmie-Young, Ian McKimmie and Barbara Joyce—and their interviews continued the theme. Strathewen had been a wonderful place to live and would be again; the community would be even stronger, more caring and more connected to the environment. This is what Michael Roth calls a "redemptive narrative", where great good can be created out of great hardship.[12] I

[12] Michael S. Roth, *Memory, Trauma and History: Essays on Living with the Past* (New York: Columbia University Press, 2012), 84.

wrote a narrative about the tree and we published both that and photographs of the tree on Collections Online.

I had seen the creation of the Poetry Tree as a positive action in the face of devastation. I was surprised and somewhat taken aback, therefore, to receive phone calls from other Strathewen residents and former residents who objected not only to our publication of the Poetry Tree material but also to the simple fact of collecting it.

I met with the women—Karen Gardam, Vicki Mitchell and Sylvia Skinner—and they told me of life in Strathewen that included sexual abuse, widespread drug use and bullying. They were angry with me for writing a story that sanitised Strathewen's past and contradicted their own experiences. By recording and publicising the poems on the tree, I and the museum were seen to be privileging the version of history they contained. Our profile of the tree, distributed to the public with the weight of the museum's reputation behind it, reinforced their feelings of helplessness and invisibility. Eventually, after many months of meetings and discussion, all three consented to be interviewed and were joined by Angela McKenzie, another former resident who had lost her house and who was even more open in her condemnation of Strathewen.

The women defied the dominant narrative of community harmony and healing to tell me their version of Strathewen. It was not easy for them, for there was a lot at stake. If you did not have a connection to the fire zone after Black Saturday, you would find it hard to appreciate the degree of pressure placed on survivors to conform to the popular narratives of resilience and recovery. Public sympathy, media attention and even recovery funding favoured individuals and communities who told a united story of grief and devastation but also of healing, hope and harmony.

The stories these women told contradicted accepted visions of recovery and opened them to criticism from other residents, especially in their frank discussions of the drug scene in Strathewen. When other residents learned I was interviewing them, I received warnings against trusting their testimony. People described them as unreliable narrators, troublemakers, even mentally ill. They wanted me to forget what I had been told.

Strathewen was frequently singled out as an exemplar of successful community-led recovery, an accolade to the recovery committee that led to increased recognition and support, but also increased scrutiny. It was possible that the publication of stories that contradicted the story of community harmony would add more stress to survivors and their efforts to cope with the present, yet failing to publish them would reinforce the women's belief that nobody listened or cared. The difficulty was compounded by the fact that the policy statement for the Bushfires Collection included the goal of "contributing to community healing".[13] Publishing their stories could threaten a fragile process of healing for some people but assist the same process for others whose stories had been generally ignored.

This was the turning point in my research, as I moved from documenting people's experiences through the fires to thinking about the relationships between people and between narratives. In the end I decided that I wanted to tell both versions of Strathewen, reporting the initial stories of the Poetry Tree but also the rebuttals to them, revisiting the original interviews to look for more complicated interpretations of the tree's symbolism, and including in the story my own biases and preoccupations.

The tension between the dominant narrative of community harmony and the defiant interviews from Karen, Vicki, Sylvia and Angela complicates and enriches our understanding of the consequences of a major disaster on community dynamics and recovery. By including the stories of those on the margins—oral history's original and perhaps still most useful purpose—we are not only validating experiences left out of official narratives but also adding new layers of interpretation to those official narratives. Far from being "best forgotten", these are important contributions to the tangled world of post-fire recovery.

Hidden Stories

In the story of the Poetry Tree, I was alerted to the contradictions between the idyll portrayed in the poems and some people's lived reality of Strathewen. You could think of them as the hidden stories of Strathewen,

[13] Museums Victoria, "Collections Policy: Victorian Bushfires Collection" (Melbourne: Museums Victoria, 2010), Policy document.

the ones that didn't make it into the fund-raising books, the newspaper articles, the dramatic re-enactments or the television specials.

Narrators, consciously or sub-consciously, suppress stories for many reasons. This is, of course, not limited to oral histories of disaster but those circumstances can make the unearthing of them particularly fraught. Talking about catastrophic events, especially one like Black Saturday that resulted in so many lives and homes lost, can bring back painful feelings that some narrators do not want to revisit. Even people who are willing to tell these stories find the process emotionally and mentally exhausting. If their stories include accounts of mistakes, bad decisions, panic or the many other things that can and do go wrong, they risk being dismissed or criticised, losing support from people close to them or from the larger community. One woman in Strathewen who spoke out against the involvement in the recovery process of people whom she regarded as criminals was ostracised and undermined by many in the community.

Survivors may be selective in their choice of stories because they are building a narrative of their lives that not only reconciles the past but also provides support for the future. The way a story is told reinforces decisions they have made, justifies actions they have taken (or not taken) and creates a structure on which they can build a place in society. Alistair Thomson calls the memories on which these narratives are based "composed memories", meaning both the process of composing a story that places difficult experiences in the framework of accepted attitudes and the striving for personal composure, creating "a safe and necessary personal coherence out of the unresolved, risky and painful pieces of past lives".[14]

There are always hidden stories. One of the tasks of the historian is to find them—and then to decide what to do with them. In the case of the Poetry Tree, I eventually decided that both sides of the story should be told. I believed that the benefits of doing so, both for the defiant narrators and for society's understanding of the challenges of bushfire recovery, outweighed the risks of community discord.

---

[14] Alistair Thomson, *Anzac Memories: Living with the Legend* (Melbourne: Oxford University Press, 1994), 10.

It is far more difficult when the risks, especially to vulnerable individuals, appear to outweigh the benefits. One woman I met through the Strathewen Community Renewal Association had lost a close family member. When I interviewed her, she gave me a concise description of events on Black Saturday in which the deceased had played a clearly heroic role. She refused to entertain any questions with regard to it. I felt that such strict adherence to a rigid version of events masked an underlying unease.

When I looked back through police statements and testimony to the Royal Commission on the 2009 Victorian Bushfires, I found a very different story: one of panic and confusion, in which the narrator had fled the fire zone, leaving the family member behind. I thought about the burden of guilt this woman might be carrying, and guessed that the recasting of the story into a heroic sacrifice was both to preserve the memory of a beloved person and to protect her own fragile composure. I was not prepared to accept the consequences of challenging such a perilously held belief. I did not tell the hidden story, but neither did I glorify the public one.

In another instance, in the gap between an introductory conversation and the recorded interview, one of the narrators altered a story about a couple who had died in the fires. He had decided to protect the remaining family members from some disturbing details about the relationship. Although I felt that the original story illustrated an important point about family dynamics in the face of a catastrophe, it was not on the record. I did use a general reference to it, stripped of identifying characteristics, to illustrate how awareness of a public audience can make someone decide to edit a narrative—in this case, out of concern for people who were already suffering a double family tragedy. I hoped that the story would add another layer to the public's understanding of the many ways in which survivors try to accommodate the needs of their community as well as their own recovery, and the way in which social pressure—even self-enforced—alters stories.

In yet another instance, the exposure of a hidden story surprised both the narrator and me. One woman described her hatred of Strathewen, yet the stories of her childhood there were full of humour and affection. When I asked her about the contradiction, she stared at me for a long moment, then out tumbled a decades-old but still painful story of sexual assault and community rejection in Strathewen, followed by a descent into drug abuse and living on the streets. I found that in our interviews, the distress of Black Saturday sometimes stirred up older pains—of previous bushfires, but also of loss of family members, of personal conflicts or of abuse. Although the revelations helped to explain the tangled relationship the woman had with Strathewen, it didn't seem to me that their inclusion in the larger, public story would either add to our understanding of Black Saturday or justify the emotional toll on the narrator.

Sometimes, the oral historian may hear a very different story from the one the narrator intends them to hear. One of my most challenging times was a series of interviews with Zelma Gartner, who survived the 1983 Ash Wednesday bushfires in Mt Macedon, Victoria. Zelma's story in many ways paralleled those of Black Saturday—a desperate dash for safety, the loss of a beloved house, a determination to rebuild—but beneath it ran a constant sub-text with two themes: the extraordinary virtues of her late husband John, and the absence of those virtues in everyone else. Zelma wanted me to see and write a straightforward account of her husband as hero, but what I saw was manipulation of the narrative in a fierce attempt to build a shrine to his memory. We did not get on. Eventually, however, Zelma's insistence on a rigid interpretation of their experiences made me question my own naïve assumptions about how survivors ought to think and behave. Her account stands in our collection as a first-person story, while my more complicated interpretation (including my education at her hands) is part of the larger narrative.

As a final example, I discovered that sometimes the stories are hidden in plain sight. Angela and Dale McKenzie lived in Strathewen in a beautiful mudbrick former church. They fled the house on Black Saturday, barely escaping with their lives, and the house was destroyed. In the aftermath, Dale made a video that included nostalgic photographs of the house before its destruction. To the casual viewer it seems that the video is a tribute to a delightful and much-mourned life in Strathewen. In fact, according to Angela's interview, the McKenzies were unhappy there, in regular conflict with neighbours and often regretting their move to the community. The video, I believe, mourned a lost dream, not a lost reality.

The Role of Difficult Narratives

After disaster, the stories that survivors tell themselves and others are important tools in the search for meaning and consolation and provide direction for the future. For Angela and Dale, the video both comforted them for what they had lost but also supported their decision not to return to Strathewen. Stories like this have a specific purpose to fulfil and the inclusion of contradictory material defeats their purpose. They are, almost by definition, incomplete and selective.

By looking carefully at the same event, in the same place, told by many different voices, I became aware of contradictions both between and within narratives but also of the importance of those same narratives in re-establishing agency and a sense of control for their narrators. Dominick LaCapra called this effort by historians to accommodate different reactions to trauma "empathic unsettlement", the combination of empathy and objectivity that places the historian in the emotional world of the narrator without losing sight of alternative interpretations and perspectives. He considered it a necessary skill in writing about traumatic events.[15]

I learned that doing disaster oral history required the acceptance that there are many different versions of 'what really happened' and the understanding that these can be in conflict with one another. I operated on the assumption that every Black Saturday story was true for that individual at that time, but not necessarily accurate or complete. Omissions or distortions were not errors or obfuscations but possible pathways into deeper insights. As Sandro Portelli has repeatedly and convincingly argued, "mistakes" can reveal important, frequently unsaid meanings: "the most precious information may lie in what the informants *hide,* and in the fact that they *do* hide it, rather than in what they *tell*".[16]

The inclusion of hidden stories in historical narrative illustrates what Portelli describes as the difference between traditional story-telling and

---

[15] Dominick LaCapra, *Writing History, Writing Trauma* (Baltimore: Johns Hopkins University Press, 2001), 78.

[16] Alessandro Portelli, "What Makes Oral History Different," in *The Oral History Reader*, ed. Robert Perks and Alistair Thomson (Abingdon: Routledge, 2006), 38.

"history-telling".[17] Alan Megill characterised it further, saying that "truthful narrative" takes account of many different viewpoints, can be justified through enquiry and analysis, acknowledges conflicting or complicating evidence and is not subordinated to a pre-established agenda.[18] The mistake I had made with the original Poetry Tree narrative was that I had published the story, not the history.

The 'difficult' stories of Black Saturday, whether openly told or hidden, are the most challenging but also, I think, the most rewarding in terms of contributing to our understanding of catastrophic events. They muddy the waters (a metaphor perhaps more appropriate to flood than to bushfire), complicating and disrupting dominant narratives. The difficult stories not only provide a different narrative but force us to re-assess the original stories for alternate meanings and consequences.

Community Healing

In exploring the complexity of disaster recovery, historians, willingly or not, become agents in community dynamics. By asking questions, analysing the poems of the Poetry Tree and publishing difficult narratives, we changed the social landscape of Strathewen, bringing into the public eye conflicting visions of what Strathewen had been and might become. We were not simply documenting social change; we had set out to participate in it, for healing is a profound social change.

The idea of a museum committing itself to 'community healing', as the collecting strategy for the Bushfires Collection states, might make many historians and curators uncomfortable. I have felt and still, to some degree, feel this discomfort. We are historians and curators, not therapists or counsellors.[19] There will also be times when the two goals of promoting

[17] Alessandro Portelli, "History-Telling and Time: An Example from Kentucky," *The Oral History Review* 20, no. 1/2 (1992): 51.

[18] Alan Megill, *Historical Knowledge, Historical Error: A Contemporary Guide to Practice* (Chicago and London: University of Chicago Press, 2007), 11.

[19] The role of historians and curators in promoting healing has been a concern for many years. Sean Field, "Beyond 'Healing': Trauma, Oral History and Regeneration," *Oral History* 34, no. 1 (Spring 2006): 31–42; James Gardner and Sarah M. Henry, "September 11 and the Mourning After: Reflections on Collecting and Interpreting the History of Tragedy," *The Public Historian* 24, no. 3

healing and documenting for the historical record will be in conflict. The concept of 'community' in this context is also problematic, because there is not in Strathewen—or perhaps in any town—a single homogenous community.

My experience in Strathewen brought out all these issues. By recording the story of the Poetry Tree, the museum increased Karen's and Vicki's distress; they had reluctantly accepted the tree's existence, but they could not tolerate the publicity generated by the museum's interest in it. By publishing their stories in turn, we exacerbated concerns within Strathewen that people's fragile resilience would be damaged. We exposed tensions and divisions within a community that was trying its best to project a positive image and confidence in an uncertain future while under intense public scrutiny.

At the same time, many people who completed oral histories with us reported a sense of satisfaction at being included in the historical record. Oral history helped some people to accommodate the events of their past by creating a narrative that provided meaning and context to their personal story. Many people also found the donation of objects and interviews to the Bushfires Collection gave them a sense of permanence and agency in a vastly changed and uncertain world. It was important for them not only to record their story but also to preserve the tangible evidence of their experience.

I spent some time working through my ambivalence about the whole idea of historians promoting healing, trying to reconcile the goals of the collecting strategy with my experience in the field. In working with the post-apartheid Truth and Reconciliation Commission in South Africa, Sean Field argued that historians should not be concerned with healing but offer "subtle support to interviewees' efforts to recompose their sense

(Summer 2002): 37–52; Kerwin Lee Klein, "On the Emergence of Memory in Historical Discourse," *Representations* 69 (Winter 2000): 127–150; Lynn Meskell and Colette Scheermeyer, "Heritage as Therapy: Set Pieces from the New South Africa," *Journal of Material Culture* 13, no. 2 (2008): 153–173; Wendy Rickard, "Oral History- 'More Dangerous Than Therapy'?: Interviewees' Reflections on Recording Traumatic or Taboo Issues," *Oral History* 26, no. 2 (Autumn 1998): 34–48; Anna Sheftel and Stacey Zembrzycki, "Only Human: A Reflection on the Ethical and Methodological Challenges of Working with 'Difficult' Stories," *The Oral History Review* 37, no. 2 (2010): 191–214.

of self and regenerate agency".[20] In the end, I restated the question as a concern with the broader society, a concern that does involve historians and history-telling:

> What [historians] can do is mix the natural, social and psychological factors of bushfire with an historically based understanding of why it happened, how it changed our view of the world and our place in it, and how it can happen again. We need to resist the comfort of a harmonising narrative that papers over the cracks. We need to examine and perhaps rethink our definitions of 'community' and 'healing' so that the emphasis is not only on the immediate response but on finding universal themes of human experience by which we can begin to understand, if not reconcile, the conflicting narratives that the Poetry Tree uncovered.[21]

This is, I think, the heart of disaster oral history—of all disaster history, actually: the delicate balancing act between different, sometimes competing, interests. It requires weighing up the benefits and the costs to the individual, the local community and the larger society. There is no neat formula to tell us what to do in specific cases, although I frequently wished for one. It comes down to making a judgement call and trying to anticipate the consequences of that judgement. Sometimes the decision is served by telling difficult stories, and sometimes it is not.

The Challenges of Telling Difficult Stories

Individual anger and community conflict are common consequences of major disasters. As psychologist Rob Gordon, who worked with many people after Black Saturday, has observed, communities will often come together in the immediate aftermath of disaster. Over time, however, old divisions often re-emerge and are joined by new frictions.[22] Sometimes these frictions have real-world consequences; I had to be aware of the

---

[20] Field, "Beyond 'Healing'", 31.

[21] Peg Fraser, "The Memorial and the Museum: Strathewen's Poetry Tree," in *Collecting the Contemporary*, ed. Owain Rhys and Zelda Baveystock (Oxford: Museums Etc, 2014).

[22] Gordon, "Community Impact of Disaster and Community Recovery".

risks of revealing difficult stories and to make sure that the narrators knew them too. One narrator was all set to name names in her allegations of organised criminal activity until I pointed out that she was risking her own safety in creating a record that could be subject to subpoena.

In trying to identify and then navigate narratives of disaster, I learned to expect the unexpected. As I found when working with the survivor of sexual abuse in Strathewen, oral histories can bring up long-forgotten memories, especially of other distressing events. Even if they do not make it into the overarching narrative, they can influence the way the story is told by the narrator and the way in which it is heard by the historian.

It was at this point that I realised that I was not just an observer; I had become a player in the process. From the objects I acquired for the museum, including poetry and textiles (two of my favourite things), to the relationships with the narrators, my research was imbued with my own subjectivity. My approach was not to try to suppress that subjectivity but to acknowledge and explore it, and to include it in the narrative. Steven High has suggested that oral history is in some ways closer to performance art than to documentary history[23]; I would add that as historians we are both in the cast and in the audience.

The commitment to a relationship between narrators and historians means an institutional commitment of time and resources. Long after our year of collecting was finished, we were still spending significant amounts of time maintaining relationships with narrators and managing the collection. When I called my book *Black Saturday: Not the End of the Story*, I was thinking about the ongoing challenges of bushfire survivors and our collective need to reassess our relationship to the cycle of catastrophic fire, but it also pertains to the continuing responsibility of institutions to maintain relationships even after the institutional goals have been met.

Many of our professional staff, including those downstream from the collecting process, were disturbed by stories of death and disaster. Acquisitions meetings were frequently gruelling, with presentations of

[23] Steven High, "Placing the Displaced Worker: Narrating Place in Deindustrializing Sturgeon Falls, Ontario," in *Placing Memory and Remembering Place in Canada*, ed. James Opp and John C. Walsh (Vancouver: UBS Press, 2010), 159–186.

objects and oral histories concerned with violent death. In describing curators' reactions to the Smithsonian's collection project after the 11 September 2001 attacks, James Gardner said that they felt "a real reticence about facing the raw emotions of the moment, a prospect outside the comfort zone of curators more at home in the world of scholarly detachment".[24] The challenge for collecting institutions is to understand the consequences of disturbing material and the necessary commitment to both the larger community and its own staff.

The most important factor in making the right decisions about difficult stories—or at least not making the wrong decisions—is time. I needed time to get to know the narrators and for them to get to know me, to develop a sense of the relationships within the community, to listen for hidden stories that might reveal unexpressed tensions or long-buried animosities, and time to figure out how to communicate to the wider public all the rich, difficult, complicated realities of life in the fire zone.

Conclusion

Oral history is a powerful tool for documenting and interpreting personal experiences through disaster. The difficult stories that emerge through oral history interviews add texture and nuance to established narratives, sometimes leading us to question the very foundations of those narratives. In the short run, difficult stories may appear to threaten individual or community recovery after disaster, but I believe that in the long term these challenging stories will contribute to a better understanding of the difficulties of survival and therefore to a stronger, more stable community recovery.

This doesn't mean that I believe that all difficult stories should be unearthed, analysed and disseminated. Stories of disaster have made me aware of the costs to narrators of reliving traumatic events and of my

[24] James Gardner, "September 11: Museums, Spontaneous Memorials, and History," in *Grassroots Memorials: The Politics of Memorializing Traumatic Death,* ed. Peter Jan Margry and Christina Sanchez-Carretero (New York: Berghahn Books, 2011), 288.

personal limitations. Like many of the narrators in this work, I have one story that will remain hidden because it is too painful for me to tell. It was horrific, although not more so than many other stories told to me by survivors, but it spoke to my own vulnerabilities and deepest fears. Even after ten years, it still awakens nightmares. It is, for me, best forgotten.

Shaped by Fire: How Bushfires Forged Migrant Environmental Understandings and Memories of Place

Gretel Evans

Jenneke was born in 1982 in the southeast of Holland and grew up surrounded by water and lush green gardens. Her family decided to migrate to Australia because her father Wytze "needed a bigger shed" and more space to expand his pottery and ceramics business. As a young child, Jenneke expected Australia would be "really hot and really dry", so she was surprised to find "it was raining" and there were "green hills everywhere" when she arrived in regional Victoria in 1989. "Surrounded by green paddocks" and dairy farms, she began to learn about her new local environment and the risk of bushfires. She joined her local volunteer fire brigade—the Country Fire Authority (CFA)—as an adult and her first call out was Saturday 7 February 2009.

That day became known as Black Saturday. The bushfires on that day remain among the worst in Victorian and Australian history; they led to

G. Evans (✉)
School of Historical and Philosophical Studies, University of Melbourne, Parkville, VIC, Australia

S. McKinnon, M. Cook (eds.), *Disasters in Australia and New Zealand*,
https://doi.org/10.1007/978-981-15-4382-1_3

the death of 173 people. Jenneke reflected on her experience afterwards and said:

> I don't think I realised what a threat [bushfires] were before actually seeing it myself [...] I knew they were an issue and I knew they existed. But I suppose I always thought they were avoidable [...] I thought you would have warning. I didn't realise how dramatic, out of control it could be.[1]

This chapter draws on oral history interviews to examine migrants' memories of bushfire in regional Victoria, their changing relationship and understanding of the natural environment, and their attachment to place. These oral history interviews reveal how the development of environmental knowledge was an important element of their early lives in Australia. This knowledge, which grew over time and often challenged pre-migration assumptions about the Australian natural environment, was critical to a growing sense of belonging in their adopted homes in regional Australia. Many interviewees' subsequent experience of bushfire as a force and power beyond human control ultimately challenged their understandings of Australia. Rather than expressing a desire to leave these hazardous locations because of their often traumatic experience of fire, many people described a new sense of attachment to place forged through the fire. Their knowledge had grown and reinforced feelings of connection and home.

The oral history interviews examined in this chapter were conducted as part of a broader project examining migrants' memories and experiences of natural disasters, especially floods and fires, in Australia. The recruitment process and self-selection of participants led to a focus on the experiences of migrants with a European background from England, Denmark, the Netherlands and New Zealand who experienced bushfires in Gippsland or the Black Saturday bushfires in areas north-east of Melbourne (Fig. 1). They ranged from 33 to 67 years of age and their diverse stories of migration included their memories of arriving as children and adults in the 1980s through to the 2000s.

---

[1] Jenneke Kijlstra-Shone, Interview with the author, Meeniyan, Victoria, 17 May 2016.

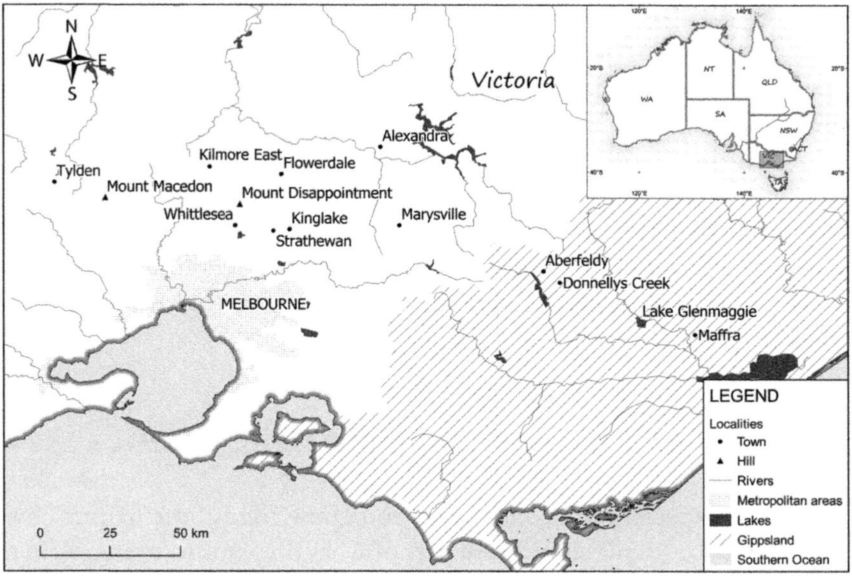

Fig. 1 Map indicating key locations in Victoria

The interviews were conducted using a life history framework. Oral historians have often turned to life history which offers a broader, holistic and nuanced approach to understanding individual memories and migration experiences.[2] Memories of key life stages, pivotal experiences and events are placed within a life history context that provides coherence and meaning to the larger life course.[3] Oral history has often focused on voices neglected or marginalised from larger historical narratives and has typically considered not just the events witnessed, but how people narrate and understand their personal experiences.

Oral histories of disaster comprise a burgeoning body of scholarship that illuminates life experiences prior to crisis and catastrophe in order to understand how individual and community memories of disasters are

[2] Ben Rogaly, "Disrupting Migration Stories: Reading Life Histories Through the Lens of Mobility and Fixity," *Environment and Planning D: Society and Space* 33, no. 3 (2015): 531; Alistair Thomson, "Moving Stories: Oral History and Migration Studies," *Oral History* 27, no. 1 (1999): 24–37.

[3] Lynn Abrams, *Oral History Theory* (Abingdon: Routledge, 2010).

shaped.[4] An important focus on rescue, recovery and rebuilding has meant the voices of people most affected by disasters have often been left out of historical records, but oral history shifts the focus and places people at the centre of the story.[5] Scholars have also highlighted the important contribution oral history offers environmental history through its capacity to illuminate the dynamic interrelationship between people, place and the natural world.[6] This discipline—its method and material—is thus well positioned to provide a detailed understanding of crisis and catastrophe, investigate migrant experiences, and usefully contribute to disaster studies and environmental history.

Before the Fire: Learning the Environment

Individual expectations, assumptions and knowledge of the natural environment prior to migration were part of what drew most interviewees to Australia, and specifically to regional Victoria. These people chose to migrate and settle in regional Australia for a sense of space and its inherent practical and aesthetic qualities. Memories of their growing knowledge, familiarity and connection with the natural environment were a key element of migrants' reflections on early settlement because it was a formative part of their initial sense of home and place in Australia. Their locale was something many people consciously and deliberately learnt about; an awareness of bushfires typically came later.

Many migrants found their previous expectations of the Australian environment were challenged by the places where they settled. Like Jenneke, Rachel was surprised by the land she encountered on her arrival in Australia in 1985 on an agricultural exchange from England. She had "pictured this land of ochre and dust and dirt and long roads and heat"

---

[4] Mark Cave, "Introduction: What Remains: Reflections on Crisis Oral History," in *Listening on the Edge: Oral History in the Aftermath of Crisis*, ed. Mark Cave and Stephen Sloan (Oxford: Oxford University Press, 2014), 1–16.

[5] Stephen Sloan, "The Fabric of Crisis: Approaching the Heart of Oral History," in *Listening on the Edge: Oral History in the Aftermath of Crisis*, ed. Mark Cave and Stephen Sloan (Oxford: Oxford University Press, 2014), 262–274.

[6] Katie Holmes and Heather Goodall, eds., *Telling Environmental Histories: Intersections of Memory, Narrative and Environment, Palgrave Studies in World Environmental History* (Cham: Palgrave Macmillan, 2017).

so was "quite disappointed" when she "ended up on [an irrigated] dairy farm in Gippsland". During her agricultural exchange Rachel met her future husband. She returned to England and they exchanged visits until "he proposed, we got married" and she emigrated in August 1988.[7]

Rachel's English agricultural background and training fostered her desire and deliberate decision to learn about the Australian flora and fauna in her new home. Her studies meant she knew the English environment well but in Australia she felt "really ignorant" because she "didn't know the wildlife ... the birds, the trees, the plant species". Rachel felt out of place because she was unfamiliar with her new local surroundings, but with the help of her mother-in-law—a very knowledgeable dairy farmer—Rachel began to start "again" with her "basic knowledge of [...] the country", and its plants and animal breeds.[8] Through learning about Australia's flora and fauna Rachel started to feel at home.

Jenneke's childhood memories of settlement revealed other ways individuals learnt about their new land and connected to place. Historian Tom Griffiths has argued that nature has too often been conceived of as a "colourful backdrop" and the "stage and setting" for more important human action, rather than something with which people actively engage or interact.[9] Jenneke's memories, however, revealed deliberate practices which actively forged early connections with place and the natural world. Contrary to Rachel's agricultural and scientific approach to learning about the natural world, Jenneke and her family "did a lot of exploring" when they arrived and went for "lots of walks and rambles". This gave Jenneke "a feeling of belonging to a place because you are familiar with your environment, with the world around you".[10] Places were made meaningful through everyday practices of walking and exploring, as well as through shared experiences like sleeping outside.[11] This sensory

[7] Rachel Dawkins, Interview with the author, Traralgon, Victoria, 3 August 2016.

[8] Ibid.

[9] Tom Griffiths, "The Nature of Culture and the Culture of Nature," in *Cultural History in Australia*, ed. Hsu-Ming Teo and Richard White (Sydney: University of Sydney Press, 2003), 67–80.

[10] Jenneke Kijlstra-Shone, Interview with the author.

[11] Peter Read, *Returning to Nothing: The Meaning of Lost Places* (Cambridge: University of Cambridge, 1996); Heather Goodall, *Rivers and Resilience: Aboriginal People on Sydney's Georges River* (Sydney: UNSW Press, 2009); Giuliana Bruno, *Atlas of Emotion: Journeys in Art, Architecture, and Film* (New York: Verso, 2007).

engagement with the natural world gave Jenneke a sense of connection to place that arguably acted as an anchor for her sense of identity and belonging into adulthood.[12]

Learning to understand regional Victoria inevitably meant developing a knowledge of bushfire risk. An important source of information and community engagement for many migrants was the Country Fire Authority. The CFA emerged from late nineteenth and early twentieth century volunteer and independent fire brigades. It was formed in 1945 and has been an important volunteer and community fire and emergency services organisation in Victoria.[13] Ian and Wytze were both motivated to join their CFA after they witnessed local outbreaks of fire. Their membership and training within the CFA gave them a technical form of bushfire education that contributed to a perception of bushfire as something to be prepared for and controlled. Ian migrated from New Zealand to Australia and had "no concept of bushfire risk" until he moved an hour north-west of Melbourne and witnessed a local bushfire in 1998.[14] Realising he was completely unprepared he immediately joined the local CFA, learnt about bushfires and made changes to his property and behaviour.

Wytze's long-term CFA involvement also informed how he perceived bushfires and how he made sense of his personal experience of the 2009 Black Saturday bushfire. Wytze and Jenneke were both members of a CFA team that was called to fight bushfires on Black Saturday. They ended up sheltering on the Marysville oval with local residents as the fire front burnt through the town. When Wytze reflected on this experience he described the speed with which the fire swept through the town and surrounding bush. People on the oval wondered if it was possible to leave and save buildings but Wytze said they were "not going anywhere because the trees were dropping left, right and centre because the eucalypts do".

---

[12] Barbara B. Brown and Douglas D. Perkins, "Disruptions in Place Attachment," in *Place Attachment*, eds. Irwin Altman and Setha M. Low (New York: Plenum Press, 1992), 279–304; David Glassberg, "Public History and the Study of Memory," *The Public Historian* 18, no. 2 (1996): 7–23.

[13] Robert Murray and Kate White, *State of Fire: A History of Volunteer Firefighting and the Country Fire Authority in Victoria* (North Melbourne: Hargreen Publishing Company, 1995); Country Fire Authority, "History Timeline," accessed 13 January 2019, https://www.cfa.vic.gov.au/about/history-timeline.

[14] Ian van der Werff, Interview with the author, Tylden, Victoria, 16 August 2016.

Wytze's CFA training gave him an understanding of fire behaviour that shaped how he responded during the bushfire. He knew that unlike the European trees of his former home, which would "creak and croak" before coming down, Australian Eucalyptus have a "very shallow root system" and "drop their branches without warning" when stressed. It was unsafe to leave the oval because these branches "weigh tonnes".[15]

Wytze's CFA involvement informed his behaviour during the fire and the way he later explained and narrated his experience. His reflections also incorporated a narrative of increasing environmental knowledge which began in Holland and developed in Australia and ultimately allowed him to protect and assist members of his local community in Victoria. In some ways, his technical knowledge of fire behaviour was disrupted by the disaster and disproved notions of fire as always manageable or controllable. Wytze commented that "I know what bushfires are, but this one really rattled our bones. No one, none of the other guys ever experienced something like that, so overwhelming".[16] He drew a distinction between what fire management practices are possible during a "normal fire" and what he and his crew encountered on Black Saturday. The fire similarly challenged Jenneke's perception of fire, formed through her involvement with the CFA as a child and adult. She wondered whether being part of the fire brigade "from such a young age" gave her "a sense of something that was able to be controlled and subdued if it needed to". It was only after Black Saturday that she realised "how dramatic, [how] out of control" bushfires could be.

The Bushfire Remembered: Sensory and Embodied Memories

A significant way migrants remembered the "elemental drama" and their personal encounter with bushfire was through embodied and sensory memories that elucidated the immediate and affective experience of fire.[17]

[15] Wytze Kylstra, Interview with the author, Marysville, Victoria, 13 July 2016.

[16] Ibid.

[17] Griffiths, "The Nature of Culture," 72.

These memories subsequently altered their behaviour on days of potential bushfire.

Hans migrated from the Netherlands to New Zealand as a teenager, and to Gippsland, Victoria, as an adult in 1996. The speed and ferocity of the 2013 Aberfeldy-Donnellys Creek fire shocked Hans who recalled how "880 acres burnt in 15 minutes. It was a big fire, it went quick". Hans knew there was "a fire in Feldy, a little fire" that could be a problem in three days' time. He went to bed but it "travelled 60 kilometres overnight" and his family were woken at 3 a.m. by a phone call and an evacuation order. Hans looked out the window and saw cars streaming away from Lake Glenmaggie, fire trucks heading towards it, and "sparks falling everywhere". It was "time to get out and go [...] it came a lot quicker than anybody thought".[18]

Hans' sensory memories of the sight and sound of bushfire were instrumental in changing his perception and understanding of fire. He saw the "hills were red behind us" and there was "a red glow and embers falling, red and blue flashing lights, people running around like headless chooks. [It was] a bit surreal". He also described the impact of the soundscape he encountered and commented that "we could hear the fire in the hills before we got evacuated".[19] According to historian Bruce Johnson, sound "produces emotional responses" or "involuntary affective interpretations".[20] When we cannot locate the source of sound—such as low-frequency sounds or sounds coming from above—it can produce emotions of anxiety and terror.[21] Hans described the "noise" of fire as inescapable and immersive and suggested that staying and fighting a fire would involve facing not just the physical flame but enduring its sound.

The bushfire challenged Hans' understandings and assumptions about the behaviour and power of fire and also threatened his sense of safety. Hans felt that they had a fire plan and were prepared, but "looking back now, well, we weren't prepared at all. Seeing what a fire can do, we weren't prepared, we thought we could fight fires, but no hope. [...] when you

[18] Hans van Wees, Interview with the author, Maffra West Upper, Victoria, 10 June 2016.

[19] Ibid.

[20] Bruce Johnson, "Sound Studies Today: Where Are We Going," in *A Cultural History of Sound, Memory and the Senses*, ed. Joy Damousi and Paula Hamilton (New York: Routledge, 2017), 16–17.

[21] Ibid., 16.

see what a fire can do, it is insignificant"; the preparations, pumps and fire-breaks felt insignificant compared to the fire. Hans felt "caught out" by the "sudden attack" of the fire. It made him feel unprepared—"this was not supposed to happen"—as they woke up to "embers falling, not standing [outside] with a hose".[22] He could not trust his eyes to know how quickly and how far the fire would travel. Its distance was not an indication of their safety.

Jenneke's personal encounter with bushfire similarly led to a changed understanding and perception of fire. Jenneke said that she "used to have an idea of how a fire would travel but then I saw one and you would never outrun it". On Black Saturday she sheltered on the Marysville oval as a CFA member as the fire front burnt through the town. She later returned to defend her home in Alexandra in case the bushfire continued to burn north. Like Hans, she no longer felt able to trust her eyes. Jenneke could "see the fire in the hills" but could not tell "how far away it was because it was just so smoky and because in my mind I was no longer sure how fast a fire would travel". Her experience the day before echoed in her mind as she walked around the house repeating to herself "there is hardly any wind, there is hardly any wind, it is not that close". She knew the wind had changed but "in my mind I couldn't quite marry those two [...] My mind was rationally saying one thing but my body was thinking flee".[23]

Sensory memories of the bushfire were embodied as part the process of remembering the fire. Similar environments and weather patterns since have often prompted those difficult memories. Jenneke said:

> I have been in windy days since then, hot windy days, and my mind is fine but my body thinks flee. I feel very tense, I feel like screaming even though I know it's fine. But it's like my body has a memory of that need to go.[24]

The fire made her "hyper aware of where I was all the time" and what the risks might be in different locations. When camping, Jenneke needs

[22] Hans van Wees, Interview with the author.

[23] Jenneke Kijlstra-Shone, Interview with the author.

[24] Ibid.

to have a fire plan in place and know where there is water before she "can feel settled in that spot". Jenneke reflected that after the fire she became more aware of her surroundings and "really aware [of] what's a fire day and what's not". She has moved "through the environment differently" because of the fire.[25]

"You Need to Respect a Fire"

Disasters like bushfires dramatically change or destroy environments, making them unfamiliar. The oral history interviews captured migrants' reactions to the environment in the aftermath of bushfire and revealed how their perception and understanding changed. Many migrants observed how nature regenerated and this was often mirrored by their own personal recovery, but a significant theme throughout multiple migrants' memories was a new sense of respect for nature that was forged through the fire.

Prior to the Aberfeldy fires Hans was aware of bushfires but "never thought about [them] until it happen[ed]". Like Jenneke and other migrants involved in the CFA, Hans expected bushfires could be fought and houses protected. He commented that "you think you can fight them but when you have to deal with the aftermath of a real fire they are totally different, they are a lot more powerful". Hans used to think disasters were "a novelty, a bit of fun" but now "I see them differently". The 2013 Gippsland bushfires challenged Hans' perception of the environment and were "a refresher call that it can burn here" even in "the middle of the irrigation district". He said he "didn't know what fire could do until I had to deal with the consequences".[26]

Hans realised that "seeing what a fire can do changes how you deal with fires". After the fire he became "more aware of what we need to do" and put "a lot more value on life than possessions. Just get out, let it burn [...] everything can be replaced except life". In the event of a future fire Hans said, "the house can just go, we would just get out and go. I am not

[25] Ibid.

[26] Hans van Wees, Interview with the author.

going to risk life and limb for a bit of brick and mortar. That is totally different to what I thought before". Previously he thought "we can risk it [...] we can deal with nature", but witnessing a bushfire changed Hans' perception of bushfires and his anticipated preparation and response during future fires.[27]

The experience also facilitated a new sense of respect and reverence for nature. Hans did not feel threatened by the environment but commented that "nature can be a bastard, usually throws two or three curved balls at once". The aftermath made him realise, "we are just so insignificant we have no chance of stopping this no hope". He has since found it difficult "to explain to people who don't live in Australia what a bushfire can do. The power of them. Awesomeness, people don't understand. You can't tame nature, you have to live with it, you can't fight it. She gives and takes, you have to accept that". His experience taught him "you need to respect a fire, you need to learn to be scared of it".[28]

After "being part of that enormous bushfire in 2009" Jenneke found it difficult to remember how she previously thought about fires. She never "expected that a bushfire could be that dramatic" or "how out of control they could be". Jenneke thought she knew what a high fire danger day meant—"oh yes, I get it"—but in hindsight she "really didn't understand what a big risk it" was. After the bushfire Jenneke became "really respectful of what fires are able to do [...] more aware of what the bush is capable of, how hot it can burn" and she felt honoured to have experienced such a life changing event.[29] This sentiment was echoed by English migrant Kat who experienced the Black Saturday bushfires in Kinglake. She witnessed "this huge force of Mother Nature" and thought it was "quite humbling [...] to have been through an experience like that". The bushfire reminded her that "the environment and Mother Nature is bigger than us and the fire is part of the cycle of life"; "we should be privileged to be here on this part of the planet". Kat felt "more respectful of the environment around" around her because now "I know what it is capable of. I still feel very at peace with it and drawn to it. It hasn't

[27] Ibid.

[28] Ibid.

[29] Jenneke Kijlstra-Shone, Interview with the author.

frightened me away from living in the bush. But I feel I have more respect for it definitely. It's mindblowing".[30]

The new sense of respect towards bushfires and nature articulated by many migrants was indicative of a changed sense of environmental understanding and knowledge borne from personal encounter and experience. The bushfire enabled them to recognise that nature is not a static backdrop or entity to be controlled but one with agency and power. Their changed understanding of the natural world did not rupture their previous relationship with the environment but in many cases renewed their attachment to place after the fires.

Attachment to Place

Migrants' new sense of understanding and respect for the environment was often articulated alongside a changed but continuing sense of connection and attachment to place. Disasters disrupt everyday life, devastate landscapes, and destroy "material traces of memory".[31] People face the disruption, destruction and loss caused by disasters in different ways. Some people decide to leave places "where they have experienced great trauma".[32] Others choose to stay and recreate or repair their attachment and connection to people and place in locations that continue to be hazardous.[33] Most migrants interviewed for this project decided to stay within bushfire-prone locations. Their oral history interviews illuminated how their relationship and connection with place before the fires continued and shaped their desire and decision to stay.[34]

---

[30] Kat Jenkins, Interview with the author, Kinglake, Victoria, 10 August 2016.

[31] Nicholas J. Entrikin, "Geographic Landscapes and Natural Disasters," in *The Place of Landscape: Concepts, Contexts, Studies*, ed. Jeff Malpas (Massachusetts: Massachusetts Institute of Technology, 2011), 114–115; Simon Swaffield, "Place, Culture and Landscape after the Christchurch Earthquake," in *Space, Place & Culture*, ed. Helen Sykes (Sydney: Future Leaders, 2013).

[32] Helen M. Cox and Colin Holmes, A., "Loss, Healing, and the Power of Place," *Human Studies* 23, no. 1 (2000): 63.

[33] Swaffield, "Place, Culture and Landscape," 149; Brown and Perkins, "Disruptions in Place Attachment," 284; Cox and Holmes, "Loss, Healing," 63.

[34] Cox and Holmes, "Loss, Healing," 63.

Jenneke said it was initially "awful to be around" Marysville straight after the fires. Being there and "seeing burnt-out things was a little bit hard", as was knowing "so many stories" of people who had died. Jenneke remembered driving through Marysville after the fires and commenting: "[it's] tidy isn't it, I wonder why it is so tidy here", before she suddenly realised: "it is tidy because everything was burnt [and] whatever was left was removed and people have started from scratch". Rebuilding efforts meant "all the houses look[ed] really new" and there was "no rubbish around the houses like there often is after years of built up farm machinery". Jenneke felt that Marysville looked "totally different"[35]:

> If you are in a town and someone chops down a tree the eye notices something is different there. In Marysville it is like that. It feels like you are walking into a different town. There are none of the visual clues that used to be there. The buildings are not there, the trees are not there. The landscape has the same undulation but that is it. And the roads are in the same spot. They are the familiar things.[36]

Jenneke felt the town was changed but it was not unrecognisable and ultimately the bushfire did not change her "want to be there or not be there". She expected that "people would leave the town because it is kind of a death trap in that way it is surrounded by forest", but although some people left, others "have been happy to rebuild and are happy that they have".[37] Peter Read's *Returning to Nothing* examined places lost to people through migration, disaster, changing land use or demolition caused by construction. These places were no longer accessible to people or had changed so much they were unrecognisable. The bushfire destroyed much of Marysville, but far from being an irretrievable or "lost place",[38] the undulation of the land and the roads provided a sense of familiarity to the town. It might be irrevocably changed, but people have rebuilt "and it seems that people are happy to be there. It feels okay".[39]

---

[35] Jenneke Kijlstra-Shone, Interview with the author.

[36] Ibid.

[37] Ibid.

[38] Read, *Returning to Nothing*; Peg Fraser, *Black Saturday: Not the End of the Story* (Clayton: Monash University Publishing, 2018), 204.

[39] Jenneke Kijlstra-Shone, Interview with the author.

Jenneke's comments were about rebuilding and recovering a sense of safety within the built and natural environments, but they also offered an insight into a continuing connection to place. Place is more than just a geographical location; it is where histories, memories, meanings and emotions are layered together. People often feel an attachment to place because they provide continuity to our lives through the physical setting, the people and activities, and the memories that are inscribed and located there.[40] Places provide an anchor for shared experiences, binding people together, providing a sense of identity, and a source of community and communality.[41] Jenneke's familiarity with the local topography, the criss-crossing of roads and her personal and shared history represented a pre-existing sense of place and an anchor during the aftermath of the fire. The bushfire challenged Jenneke's sense of safety and how comfortable she felt in Marysville during and after the fires, but her previous attachment to place meant her bushfire memories were ultimately woven into this location, which contributed to a continuing connection and relationship with this place.

Attachment to place arguably contributed to some people's desire and commitment to stay and rebuild. Kat's family returned to Kinglake after the fires and began to rebuild. In order to "deal with our trauma" they clung "to the plans that we had before the fires" and "threw ourselves into getting the house built and moving into the house". Building a greenhouse and shed and growing a garden were activities that were "a great distraction for us".[42] There are political and economic factors that contribute to people's decision to rebuild, but there are also emotional values that encourage people "to stay in an environmentally vulnerable place".[43] Doreen Massey has described place as "an attempt to find rooted security, to establish fixed identities amidst—and in opposition to—the flux,

[40] Swaffield, "Place, Culture and Landscape," 144.

[41] Mike Crang, *Cultural Geography* (London: Routledge, 1998), 22, 23; Brown and Perkins, "Disruptions in Place Attachment," 280.

[42] Kat Jenkins, Interview with the author.

[43] Eleonora Rohland et al., "Woven Together: Attachment to Place in the Aftermath of Disaster, Perspectives from Four Continents," in *Listening on the Edge: Oral History in the Aftermath of Crisis*, eds. Mark Cave and Stephen Sloan (Oxford: Oxford University Press, 2014), 193.

fragmentation, and disruption of a postmodern world".[44] After the bushfire's disruption to their lives, rebuilding arguably gave Kat and her family a sense of order and control that also deepened their sense of attachment to place.[45] Kat reflected that they had put "so much time and energy" into rebuilding their home that there was "no way we could ever move […] the idea of moving somewhere else and leaving that all behind" was unthinkable.[46] Their decision to rebuild their house represented an economic and an emotional investment in this location.

Staying in Kinglake ultimately generated new memories and new connections. Despite the traumatic experiences of the bushfire Kat still saw "a really beautiful side to Kinglake and I feel really connected to it, the earth and the environment here".[47] Kat participated in a fund-raising walk and trained by walking through "the bush and up mountains and all around the local scenery […] it is just the most amazing place. It's so beautiful and in that respect I adore being here. I love being here, it is a remarkable space to be […] It definitely feels like home".[48] Echoing back to migrants' memories of learning about their locale before the fire, Kat's memories of walking through the local environment demonstrated how she reconnected after the fires and rekindled a sense of attachment to the place she chose to call home.

Etchings: Embedded Memory

Places are embedded with layers of history, memory, shared experiences and meaning. They are a "bank of cultural memories" where residues of the past are embedded in the landscape and written over, but not erased.[49] Memory attaches itself to sites, and this was evident within migrants' oral

[44] Shelley Trower, "Introduction," in *Place, Writing, and Voice in Oral History*, ed. Shelley Trower (New York: Palgrave Macmillan, 2011), 13.

[45] Rohland et al., "Woven Together," 187.

[46] Kat Jenkins, Interview with the author.

[47] Ibid.

[48] Ibid.

[49] Crang, *Cultural Geography*, 22, 23.

history interviews.[50] Bende grew up in Denmark and migrated permanently to Australia after she met her husband in Melbourne on a working holiday. She chose to live in the country because "if you had to migrate to another country you would want to live in the country".[51] On Black Saturday the Kilmore East bushfire initially travelled in a south-easterly direction and was heading towards the Whittlesea area before a wind change caused the fire to burn towards towns including Kinglake and Flowerdale. That wind change likely kept Bende's house safe but it led to the death of her parents-in-law in Flowerdale. Bende's family eventually sold their property once they were ready to part with it and the trees had grown green from burnt black. They moved away to be safe from fires but stayed within a similar area to be close to friends and have easier access to town. Bende was initially drawn to the space in Australia where she could "look into the distance, there is nothing that is breaking my view". The floor to ceiling windows in their new house provided a beautiful view of their property and of Mount Disappointment.[52] Mount Disappointment formed the backdrop to our interview on a sunny winter's day. The skeletons of burnt trees were visible in the distance.

Memories of the bushfire have become embedded in this place. Mount Disappointment is "a keeper of memories" that offered a unique way to remember the bushfires and the people Bende lost in the fires.[53] Bende said:

> When I sit here and look out the window I think about Grandpa. But I feel like I have to, that is a part of my everyday now. As I drive to work I drive past all the burnt bits. It is one of those things. I think we did this and this and this and maybe could have done that.[54]

This was not the place where her father-in-law died and this was not the house Bende lived in during the bushfires, but this place and its view of the mountain became an important focus for her memories. Maria

[50] Pierre Nora, "Between Memory and History: Les Lieux De Mémoire," *Representations* 26 (1989): 22.

[51] Bende (last name withheld), Interview with the author, Eden Park, Victoria, 2 August 2016.

[52] The explorers Hume and Hovell named Mount Disappointment in 1824 when they were disappointed the dense vegetation blocked their view of Port Phillip Bay.

[53] Edward S. Casey quoted in: Maria Tumarkin, *Traumascapes: The Power and Fate of Places Transformed by Tragedy* (Melbourne: Melbourne University Press, 2005), 129.

[54] Bende, Interview with the author.

Tumarkin has observed that internal processes of thinking, remembering and grieving need "the right kinds of places to unfold" and traumascapes "were often these kinds of spaces".[55] They offer a way to explore the way specific places can be embedded with histories and memories of trauma, violence and loss.[56] Mount Disappointment is a type of traumascape with unique characteristics due to the layering of environmental and human histories and personal memories, but it nonetheless provides an important insight into how experiences of loss can be bound to physical places. These places are an important part of remembering the fires. Bende reflected that people "deal with grief in different ways". Other people "don't think about the fires everyday like we do here because we see the mountains […] they can get away from it whereas we can't get away from it and we don't want to get away from it".[57] This particular place provided both an opportunity and an obligation to remember. Mount Disappointment was a personal place of memory embedded with Bende's memories of the fire.

> Do you still think about the fire regularly?
>
> Yes, but not to the extent where I sit and cry about it. It is like I said, if I put the kettle on I look out and I think yes, I wonder what that looks like up there […] I wonder how big the bushes have got since the fires. And I stand here at the [kitchen] sink looking and thinking when the people were here—because we weren't here—they would have had the best view to see where the fire was going. I think of that quite often.[58]

Conclusion

The bushfires recalled by migrants did not just change the physical environment but also changed their individual perceptions, understandings, and relationship with place and the natural world. The oral history interviews provided an insight into migrants' understanding and perception of Australia prior to their experience of bushfire, which subsequently

---

[55] Tumarkin, *Traumascapes*, 36.

[56] Ibid.

[57] Bende, Interview with the author.

[58] Ibid.

changed how they related to their surroundings and saw their place in the world. Migrants' personal experience of bushfire altered their relationship with the natural environment, and in many cases, deepened their attachment and connection to place. Their changed relationship demonstrated the importance of personal experience and memory as a way to understand how people create meaning from their experiences and choose to live in these places.

Their memories demonstrated how as newcomers they learnt about the land and formed their perceptions, assumptions and knowledge of Australia's flora, fauna and fire. The bushfires, however, were not something that could be controlled, but an active force that changed their understanding of the environment and relationship to place. Their personal experiences, captured through sensory and embodied memories, began to shape how they moved through the landscape after the fires. In some cases it confirmed existing technical knowledge of fires, but most often the fires led to a new respect and understanding of bushfire and the natural world. This contributed to a new sense of connection with place as an important site of meaning, memory and mourning.

The bushfire challenged migrants' perceptions of the environment and opened their eyes to the risk and reality of future fires. Knowing that fire is not always manageable or controllable was an important feature within migrants' ongoing process of learning about the Australian environment. Their ability to understand and respond to the environment was developed through the fire and was interwoven with a commonly continuing sense of belonging and attachment to place. These oral history interviews provide evidence of an evolving relationship with the environment. Despite—or because of—the destruction brought by fires, these bushfire-prone places provided people with a way to remember and mourn and gave many migrants a sense of attachment and connection to the places they had chosen to call home.

Acknowledgements I would like to thank Angus Ferguson, Emily Fitzgerald, Alessandro Antonello and the fellow contributors to this book for their suggestions and feedback on this chapter. This research was supported by an Australian Government Research Training Program (RTP) Scholarship.

Placing Memories of Unforgettable Fires: Official Commemoration and Community Recovery After the 2003 ACT Firestorm

Scott McKinnon

On January 18, 2006, in Stromlo Forest Park on the western outskirts of Canberra, a ceremony was held to formally open the Australian Capital Territory (ACT) Bushfire Memorial. Four people had been killed and nearly five hundred homes destroyed in the fires that had swept through the ACT on a stiflingly hot day exactly three years before. In contrast, the day of the opening ceremony was cold and wet. Speaking to the assembled crowd of around 200 survivors, officials and locals huddled under umbrellas, survivor Jane Smyth asked, "We don't need to be reminded about the effects of weather on our lives, but if rain had fallen like this on January 18th 2003, where would you be today?"[1] ACT Chief Minister Jon Stanhope officially dedicated the government-funded site and

[1] "The blackest of days, 3 years ago, remembered," *Canberra Times*, 19 January 2006, 1.

S. McKinnon (✉)
University of Wollongong, Wollongong, NSW, Australia
e-mail: scottmck@uow.edu.au

S. McKinnon, M. Cook (eds.), *Disasters in Australia and New Zealand*,
https://doi.org/10.1007/978-981-15-4382-1_4

declared, "The Bushfire Memorial is for the people of Canberra. It tells the story of their ordeal and of their resilience."[2]

Just down Cotter Road on the route to the memorial, Peter and Jenni Farrell had parked their car and stood with a homemade sign that read "Shame Jon Shame." The Farrells had managed to save their home during the fires, but remained angry at the lack of warning given to Canberra residents. Jenni Farrell told the *Canberra Times*, "Peter and I were prepared ourselves. We knew that this was going to happen. Thousands of other people didn't know and it was left to the last minute when there were people in their shorts and thongs with saucepans trying to put out their homes."[3] The Farrells saw Jon Stanhope's government as complicit in the disaster and objected to his presence at the ceremony. They were not alone. Other survivors boycotted the event, and some planned an alternate ceremony at the new memorial some weeks later.

This chapter examines the first three years following the 2003 ACT firestorm, focussing on the intertwining of commemoration and recovery in anniversary events and memorial sites. Kroepsch et al. describe disaster anniversaries as "key periods for negotiation of an event's causes, its meanings, and its implications in media."[4] Through oral history interviews conducted by Dr Mary Hutchison for the National Library of Australia between 2004 and 2008, as well as newspaper reports and government documents, I examine each anniversary as a snapshot in time, through which to trace both the significance—and the substantial challenges—of commemoration as an element of recovery processes.

The history of commemorative acts and sites offers a valuable entry point into understanding disasters not as temporally discrete moments, but as enduring events with social, political, cultural and environmental impacts that ripple through time. Historian Rebecca Jones has labelled droughts "slow catastrophes,"[5] an apt description for that form of disaster

---

[2] "Bushfire memorial unveiled," *Chronicle*, 26 January 2006, 5.

[3] "Weston couple stage their own protest, down the road," *Canberra Times*, 19 January 2006, 2.

[4] Adrianne Kroepsch, Elizabeth A. Koebele, Deserai A. Crow, John Berggren, Juhi Huda, and Lydia A. Lawhon, "Remembering the Past, Anticipating the Future: Community Learning and Adaptation Discourse in Media Commemorations of Catastrophic Wildfires in Colorado." *Environmental Communication* 12, no. 1 (2018): 136.

[5] Rebecca Jones, *Slow Catastrophes: Living with Drought in Australia* (Clayton, Vic: Monash University Publishing).

given its enduring and indefinable timeframes. And yet there is also an argument to be made that all disasters are "slow," particularly when viewed, in the words of Sara Le Menestrel, through the "temporality of disaster as process as opposed to an event."[6] The ACT firestorm swept across bushland, farms and suburbs with frightening speed, destroying homes and taking lives in a matter of moments. The history of the disaster didn't end with the extinguishing of the final flame, however, nor did it end with the conclusion of the final government inquiry or the rebuilding of the final home. The disaster endures in memory, and the ways in which the disaster is (and has been) remembered have lasting consequences both for the lives of survivors and for how the next fire will be experienced.

Historian Tom Griffiths describes bushfire in Australia as "a recurrent nightmare. We know this phenomenon, we know the specific contours of the event, and we even know how people live and how people die."[7] Amid the escalating impacts of climate change, it remains deeply concerning that, according to Griffiths, "we still have not come to terms with what we have already experienced."[8] It is in part through collective acts of memorialisation that communities attempt to "come to terms" with a disaster. Better understanding the forms of memory-making that take place might help us understand why bushfires are so often remembered as isolated tragedies safely contained in the past, rather than as cyclical elements of life in large areas of Australia. As times move on and communities change, the urgency of recent memories fades, relegating past fires to history. And yet, in the words of historian Stephen Pyne, "damaging fires will come, they will spread as the landscape allows, and they will inflict damage as structures permit."[9]

---

[6] Sara Le Menestrel. 2014. "Memory Lives in New Orleans: The Process and Politics of Commemoration," in *Hurricane Katrina in Transatlantic Perspective*, ed. R. Huret and R. J. Sparks (Baton Rouge, LA: LSU Press, 2014), 4.

[7] Tom Griffiths, 2009. "'An Unnatural Disaster'? Remembering and Forgetting Bushfire," *History Australia* 6, no. 2 (2009): 35.2.

[8] Ibid.

[9] Stephen J. Pyne, "Historian Stephen J. Pyne on the Australian Fires." *Peeling Back the Bark: Exploring the Collections, Acquisitions, and Treasures of the Forest History Society* (Forest History Society, 2019). https://fhsarchives.wordpress.com/2009/02/10/historian-stephen-j-pyne-on-the-australian-fires/.

Memorials and commemorative events play a number of important emotional, political and educational roles after a disaster. Ideally, disaster commemoration will achieve two outcomes. First, it will contribute to the recovery processes of disaster-impacted communities. Commemoration and reflection can comprise valuable acts of healing for survivors. Second, commemoration may serve as a reminder that future disasters are likely and that the need to prepare is essential. Garde-Hansen et al. illustrate the need for "active remembering"[10] in disaster-prone communities, through which local knowledge and experience is drawn on to maintain awareness and reduce disaster risk.

Achieving each of these aims is highly challenging. No form of commemoration will ever encompass the complexities of an event that was experienced—and is remembered—differently across a diverse population.[11] The memories of the Farrell family, for example, could not be reconciled with an official commemorative process that ignored their concerns about perceived government failure. Commemoration can be an act of healing, but it can also be politically contentious and harmful in ways which exacerbate pain and trauma. Equally, commemorative rhetoric potentially positions disasters as successfully managed historic events, thus limiting opportunities to revise flawed policies, educate communities and encourage disaster preparation.[12]

Below, I examine the events of 2003 and the following three anniversaries, exploring the forms of commemoration employed by survivors and the ACT Government, including commemorative events and sites. Outside the scope of this chapter are the smaller, although no less important, forms of commemoration that took place in homes and neighbourhoods, developed by survivors themselves. Ultimately, I argue that the written history of disasters should incorporate slow recoveries and the attempts, both formal and informal, to maintain memories of the event. As argued by Ashton *et al*, "There is not just a history of memorials

[10] Joanne Garde-Hansen, Lindsey McEwen, Andrew Holmes, and Owain Jones, "Sustainable flood memory: remembering as resilience," *Memory Studies* 10, no. 4 (2017): 390.

[11] Susan Nicholls, "Disaster memorials as government communication," *Australian Journal of Emergency Management* 21, no. 4 (2006): 36–43.

[12] Scott McKinnon, "Remembering and Forgetting 1974: The 2011 Queensland Floods and memories of an earlier disaster." *Geographical Research* 57, no. 2 (2019): 204–214.

as objects; there is at the same time a history of memory and commemoration."[13] Paying attention to these histories allows us to understand, first, the complexity of remembering as an element of recovery and, second, the enduring meanings of past bushfires in communities destined to face more fires in the future.

The 2003 ACT Bushfires

Located on Ngunnawal country, Canberra was established in the early 1900s as the purpose-built capital for a newly federated nation (Fig. 1). Often referred to as the "Bush Capital," many of Canberra's leafy suburbs were developed with close connections to nearby bushland and to the pine plantations first planted in the 1920s. This created an attractive

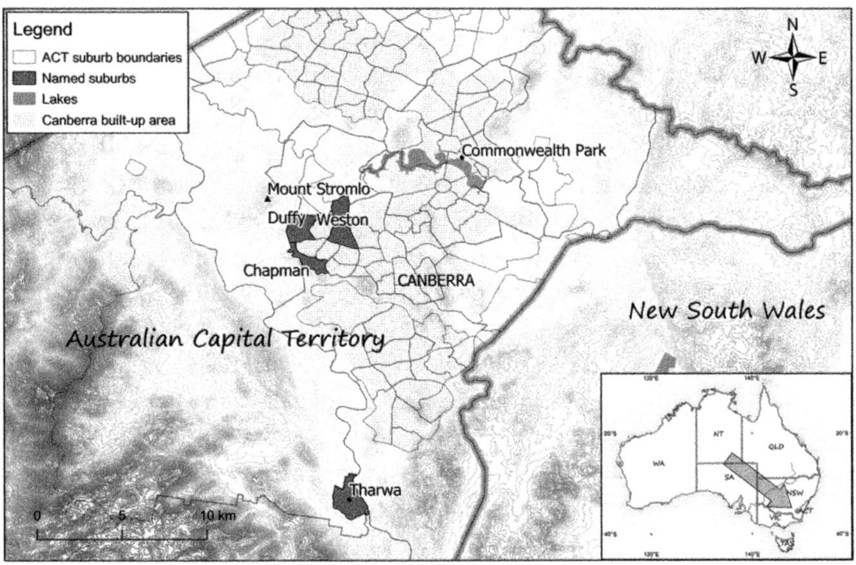

Fig. 1 Map indicating key locations in the ACT

[13] Paul Ashton, Paula Hamilton, and Rose Searby, *Places of the Heart: Memorials in Australia* (North Melbourne, Vic: Australian Scholarly Publishing, 2012), 12.

living environment blurring the line between nature and the city. It also ensured greater fire risk.

Canberra sits within the ACT, an area of 2400 square kilometres around 70 per cent of which, in 2003, was comprised of bushland and lumber plantations. From 2001 onwards, the ACT and surrounding areas of New South Wales (NSW) had endured one of the most severe and prolonged droughts in Australia's recorded history, resulting in extremely dry conditions in eucalyptus forests, open grassland and pine plantations.[14] Fire risk was exacerbated that January by record temperatures and below normal humidity and cloudiness.

Separate fires were started by lightning strikes in the hills to the west and south-west of Canberra on January 8, 2003.[15] Crews were dispatched by ACT and NSW fire services; however, the fires were hard to access and difficult to fight. The fires burnt for eight days, before joining up and moving towards the Canberra suburbs on January 17.

On January 18, residents in Canberra woke to a hot day that would peak in the forties. As the day progressed, smoke and ash filled the sky, which darkened to ominous reds and blacks. In an oral history interview, Jane Smyth remembered, "It was extremely hot and very dry. And, I don't think it's just with hindsight I've decided this. I think it was—there was an eerie feeling about it."[16] Still, there was no official warning of danger. Although authorities had issued statements about the fires throughout the previous week, they did not provide any information indicating the threat to Canberra itself. The first such warning would not come until 2.40 p.m. Coroner Maria Doogan described this message as "too little, and it was delivered far too late."[17] By 3 p.m., several suburbs were on fire.

In a little more than a week, the fire had burned across rural and bush communities, farmland and suburban homes. The people fleeing or fighting the fires over that time had diverse understandings of and experiences with bushfire. Some had been listening to the radio for updates, preparing

[14] Nicholls, "Disaster memorials as government communication".

[15] Ron McLeod, R. 2003. *Inquiry into the Operational Response to the January 2003 Bushfires in the ACT* (Canberra: ACT Government, 2003), 15.

[16] Jane Smyth, Interview with Mary Hutchison, Canberra Bushfires 2003 Oral History Project, National Library of Australia, 2004.

[17] Maria Doogan, *The Canberra Firestorm: Inquests and Inquiry into Four Deaths and Four Fires Between 8 and 18 January 2003: Volume II* (Canberra, ACT: ACT Coroners Court, 2006), 159.

their homes and properties and putting fire plans into action. Others had no knowledge that their homes could be threatened, had no plan that they could act on and only realised they needed to take action when the fire reached their suburb. All faced a terrifying landscape of heat, roaring wind and smoke that blacked out the sun. Pat Gregory remembered, "The noise was the most horrific experience. You can't explain it, you'd have to be in it."[18] Sophie Penkethman was just 13 years old and recalled, "So me and my mum both had scarves around our faces and we were hosing down the house … And by then the sky was like turning black, so we were going, yes alright maybe it will happen."[19] Barbara Pamphilon recalled, "[W]e all see on television these idiots out fighting fires without proper clothes on. … Well, hello, here's one of them, me. It hit so fast."[20]

By early evening, 488 homes had been destroyed in suburban and rural areas.[21] Many more were damaged. Four people—Dorothy McGrath, Alison Tener, Peter Brook and Douglas Fraser—had died. Three more were severely burned. More than 5000 people had been evacuated to emergency shelters and many more were sheltering with family or friends. Many thousands of animals perished, including pets and livestock. Through the course of the fires, an area of 160,000 hectares had burnt, including 31,000 hectares of rural leases. The pine plantation previously situated on Mt Stromlo was now mostly gone. It was there that the memorial would be built three years later.

2004: The First Anniversary

One year on from the firestorm, the transformed landscape of Canberra suburbs like Duffy, Chapman and Weston ensured that the fires remained unforgettable. In Chauvel Circle, a Chapman street labelled "the

---

[18] Pat Gregory, Interview with Mary Hutchison, Canberra Bushfires 2003 Oral History Project, National Library of Australia, 2004.

[19] Sophie Penkethman, Interview with Mary Hutchison, Canberra Bushfires 2003 Oral History Project, National Library of Australia, 2004.

[20] Barbara Pamphilon, Interview with Mary Hutchison, Canberra Bushfires 2003 Oral History Project, National Library of Australia, 2004.

[21] Gail Winkworth, Christine Healy, Merrilyn Woodward, and Peter Camilleri, "Community capacity building: Learning from the 2003 Canberra bushfires," *The Australian Journal of Emergency Management* 24, No. 2 (2009): 6.

symbolic ground zero of Canberra's bushfire disaster"[22] by Melbourne newspaper the *Age*, 19 of 23 homes had been destroyed. By January 2004, rebuilding had progressed on only five. For the residents whose homes had survived, life in the street was highly challenging. Eda McGloughlin told the *Age*, "It's been a very traumatic year. We're living in a very desolate environment."[23] For residents like Eda, to step out the front door of home was to be bombarded with memories of fire in the lingering smells of ash and burnt materials, the empty blocks where houses had once stood, the missing trees and the absence of birdsong.

Physical, mental and emotional recovery for survivors remained an ongoing and difficult process. In June 2004, Alan Latta recalled, "Even now, I get very, um, shaky. I weep, I weep a lot. I can be driving down the road and it will just come out for no reason. I get nervy. I feel like I'm shattered."[24] Individuals who have experienced a disaster often describe embodied forms of memory which revive fear and trauma.[25] For bushfire survivors, the smell of smoke can cause a flash of panic. Equally, material objects might prompt difficult memories. Jane Smyth's home in Chauvel Circle had been completely destroyed. She remembered, "Even now, [my husband] will say something like, 'Do you have a pencil sharpener?' and I'll say, 'No.' I'll say, 'I had one.' 'Oh, it must be at Chapman,' we'll say. You know, because we don't want to say, 'Oh, we lost that too.'"[26] Alan Latta's home and its contents were also lost to the flames. In 2004 he was self-employed as a cleaner, working in other people's homes filled with familiar belongings. He stated, "You think, oh geez I used to have one of these. So you can't get away from it. You don't have to talk about it because it's always there."[27]

---

22 Dan Silkstone, "Rising from the Ashes," *The Age*, 17 January 2004, 3.

23 Ibid.

24 Alan Latta, Interview with Mary Hutchison, Canberra Bushfires 2003 Oral History Project, National Library of Australia, 2004.

25 Gail Adams-Hutcheson, "Embodied Vibrations: Disastrous Mobilities in Relocation from the Christchurch Earthquakes, Aotearoa New Zealand," *Transfers* 7, no. 3 (2017): 3–37; see also Gretel Evans, this volume.

26 Jane Smyth, Interview with Mary Hutchison, Canberra Bushfires 2003 Oral History Project, National Library of Australia, 2004.

27 Latta, Interview with Mary Hutchison.

In the evocative phrase of historian Peter Read, "Memories are ghosts that won't lie down."[28] For some, formally commemorating the first anniversary of the bushfires seemed unnecessary or even odd when they lived surrounded by the disaster every day. Complicating the value of commemoration as an act of healing is the fact that many survivors of the 2003 fires were actively trying to forget past traumatic events. For these survivors, commemoration served as an unwanted reminder and became a perceived hindrance, limiting their progression towards a future beyond the fires. Asked whether the idea of anniversary events interested her, Pat Gregory replied, "No, no it doesn't. No, it doesn't appeal to me. I think I'd rather move on and you know, a bit like everything else, sort of, you have to get over it and move on."[29]

Pat's expressed desire to "move on" was constrained by the residual anger she carried about the management of the fires. She had endured a terrifying experience on her rural property and her brother-in-law's home, on a neighbouring farm, had been destroyed. Pat identified specific emergency management decisions she believed had left their properties vulnerable. It is possible that any form of official commemoration would, for Pat, be too interwoven with ongoing anger to offer any positive benefit. Journalist Jordan Flaherty, describing the first anniversary of another disaster, has asked, "How do you commemorate the anniversary of something that is still happening?"[30] For many people in the ACT, the disaster endured in ways which complicated the value of commemoration.

Others, however, saw marking the first anniversary as valuable and necessary. Around 3000 people attended a commemorative event organised by the ACT Government and held at Canberra's Commonwealth Park. At the event, a range of speakers and musical performances reflected on the fires and on the recovery process. According to the newsletter of the ACT Bushfire Taskforce, a government-run support agency, "Common to each of the performances was a feeling of hope and optimism for the future, and the sense of community which has emerged in the aftermath of the

---

[28] Read, Peter. *Returning to Nothing: The Meaning of Lost Places* (Cambridge, UK: University of Cambridge Press, 1996), 111.

[29] Gregory, Interview with Mary Hutchison.

[30] Quoted in Le Menestrel, "Memory Lives in New Orleans": 4.

fires."[31] The event was thus configured as a combination of past, present and future. While acknowledging the trauma of the disaster and the lives lost, the commemoration offered reassurance of happier times ahead and celebrated the community formed by the bushfires.

Memorial events and sites are often constructed as a celebration of resilience. For some survivors, this form of commemorative framing is of significant value. Narratives which highlight community resilience carry assurance that recovery is possible. The ACT Taskforce newsletter thus reported, "During the final performance of [the song] *Rise Up*, the audience stood as one—a finale which demonstrated the strength, courage and optimism of the community."[32]

This celebration of community had particular currency for some Canberra residents. In news media representations, Canberra is often depicted as a characterless city of politicians and bureaucrats without the sense of heart and community that bonds other regional cities of similar size. After the fires, a new narrative emerged in media reports and political rhetoric: the strong and supportive response of Canberra residents to the needs of bushfire survivors had transformed the soulless capital into a caring community. In his speech at Commonwealth Park, Chief Minister Jon Stanhope declared, "Like a native seed, awaiting the heat of fire to trigger its germination, our city, our community, has seen the germination and growth of its true soul."[33]

Alan Latta and his family had moved to Canberra 12 months before the fires and he had struggled to make connections to the local community. The fires transformed his feelings of connection to the city and its people. He stated, "I always thought Canberra was a mean town, but it's not. I will think of Canberra as a caring town, which it is now, since that fire, Canberra is a caring town."[34] People often respond to disasters with extraordinary bravery and acts of selfless generosity. Rebecca Solnit argues, "Horrible in itself, disaster is sometimes a back door into paradise, the paradise at least in which we … are each our sister's and brother's

[31] "ACT Bushfire Recovery Taskforce," *Community Update* 47 (2004): 1.

[32] "ACT Bushfire Recovery Taskforce": 1.

[33] "ACT Bushfire Recovery Taskforce": 2.

[34] Latta, Interview with Mary Hutchison.

keeper."[35] For Alan, community, gratitude and acknowledgement were interwoven in his decision to attend the anniversary event. He recalled, "The whole day was just a thank you day from my side. That's all I wanted to do was go around and thank everybody for all the help and everything they did for us ... It was a good day."[36]

Some survivors were resistant, however, to narratives which identified a new-found soul emerging through the disaster. Emma Walter was also at the anniversary event, which she found important and valuable. She remembered, however, being unhappy with speeches from politicians declaring a new community. She stated, "Canberra's always had a soul, it's always had a heart ... I was a bit shocked when they said, 'Now everyone else knows' and I thought, I don't care what everyone else knows. ... I do agree that the community spirit that came out of it was amazing but, I mean, it's always been there."[37] Barbara Pamphilon identified similar rhetoric in news media accounts of the fires. While she was pleased that the rest of Australia might now understand Canberra as a place of "real people" and not just "heartless bureaucrats," she argued that narratives declaring that "Canberra has now found its soul" ignored important issues. "I think we're getting a lot of the slick predictable stuff. And because I'm living the messy, unpredictable, chaotic sort of stuff reaction, I don't tolerate well the platitudes and the quick and easy take home lines."[38]

Barbara identified the possibility that commemorative rhetoric praising a strong, heroic community response might fail to recognise the difficulties of dealing with catastrophe. She expressed a desire to see "some decent documentaries done that were much more investigative and probing." Commemorative processes which offered a uniformly positive narrative of resilience provided vital encouragement to a traumatised community struggling to recover. These narratives risked, however, limiting possibilities for identifying vulnerability and for reforming policy and practice into the future.

---

[35] Rebecca Solnit, *A Paradise Built in Hell: The extraordinary communities that arise in disaster* (New York, NY: Penguin, 2009), 5.

[36] Latta, Interview with Mary Hutchison.

[37] Emma Walter, Interview with Mary Hutchison, Canberra Bushfires 2003 Oral History Project, National Library of Australia, 2004.

[38] Pamphilon, Interview with Mary Hutchison.

2005: The Second Anniversary

An editorial in the *Canberra Times* described the second anniversary as "a moment to take a pulse check on how our community has recovered from the fires, what further support residents may require and to allow another opportunity … to pause and reflect."[39] The editor stressed that the community's recovery was still very much an ongoing process, noting that "many people remain traumatised by events." The article pointed out two significant factors that continued to produce anger and anxiety among the disaster-impacted community: first, an ongoing coronial inquest into the fires; and, second, the slow rebuilding process.

Two years on, a third of the homes destroyed by the fires had been rebuilt and re-occupied, leaving hundreds of lots sitting empty and hundreds of people yet to re-establish a permanent home.[40] One hundred and fifty-six blocks had been sold, suggesting that many former residents had decided to move elsewhere. With so many houses being rebuilt or repaired in the city, building costs had sky-rocketed and some Canberra residents could not afford to begin rebuilding their home, particularly those who had been under-insured. The financial impacts of the fires were enormous and debilitating.

Official commemorative events on the second anniversary were of a much smaller scale than the first. A "twilight ceremony"—including a sausage sizzle and musical performances—was held at the Weston Creek Community Primary School, organised by the Weston Creek Community Council.[41] A crowd of around 300 people attended, in contrast to the 3000 at the Commonwealth Park event 12 months before. Perhaps of more significance, the ACT Government marked the anniversary by announcing that a site had been chosen for an official fire memorial. Acting Chief Minister Ted Quinlan stated that the memorial would represent "both the devastation of the fires and the reflective journey of recovery."[42]

---

[39] Editorial. "To serve the national city and through it the nation: Anniversary a day to reflect," *Canberra Times*, 18 January 2005, 8.

[40] "Firestorm's grip of torment," *Canberra Times*, 15 January 2005.

[41] "Community to remember 2003 bushfires," *ABC News*, 18 January 2005, https://www.abc.net.au/news/2005-01-18/community-to-remember-2003-bushfires/620612.

[42] Doherty, Megan. 2005. "Memorial to mark healing: Design to be unveiled on fire's anniversary." *Canberra Times*, 18 January, p. 1.

The announcement of the memorial site came at the end of an extensive consultation process co-ordinated by the ACT Government arts funding body artsACT and the Department of Urban Services. Susan Nicholls argues that the ACT Government "conducted the development of the memorial with some finesse … It was assiduous in both consultation and transparency."[43] An advisory committee comprised of bushfire survivors and government representatives was formed and community views were gathered over several months. The level of active community participation likely prompted feelings of ownership and participation in the finished memorial. Significantly, the process was undertaken in an atmosphere of ongoing anger at the ACT Government from many survivors. Rather than dissipating over time, this anger intensified around controversies related to the coronial inquest.

Coronial hearings into the causes of deaths in the fires had begun in June 2003, headed by Coroner Maria Doogan. At the time of the second anniversary, the inquiry was far from complete. Indeed, Chief Minister Jon Stanhope and other government ministers had taken the Coroner to court, claiming that she should remove herself from the inquiry because of perceived bias.[44] Doogan refused to step down and the Supreme Court case took months to complete. The Court found that Doogan did not need to recuse herself, and her final report was issued in December 2006.[45]

Many ACT residents had hoped the inquiry would address their concerns about the management of the fires, seeing the hearings as a chance to uncover truths previously hidden. Their anger at, and suspicion of, Jon Stanhope was only exacerbated by the court case against Doogan. This anger led some in the community to reject the idea of an official memorial altogether. In an oral history interview in 2005 Val Jeffery, a community leader in the small town of Tharwa and a volunteer fire-fighter over many decades, stated, "the best memorial they could give us was to tell us the truth."[46]

---

[43] Nicholls, op. cit., 42.

[44] "Bushfire inquest legal wrangle continues," *ABC News Online*, 23 May 2005, https://www.abc.net.au/news/2005-05-23/bushfire-inquest-legal-wrangle-continues/1575990.

[45] Maria Doogan, *The Canberra Firestorm: Inquests and Inquiry into Four Deaths and Four Fires between 8 and 18 January 2003: Volumes I and II* (Canberra, ACT: ACT Coroners Court, 2006).

[46] Val Jeffery, Interview with Mary Hutchison, Canberra Bushfires 2003 Oral History Project, National Library of Australia, 2005.

A "Community Consultation Discussion Paper" on the proposed memorial was issued by artsACT in June 2004. The paper acknowledged the objections received from some members of the community, but reported overwhelming support for the idea of a memorial. The issue of anger was specifically addressed: "Anger is a legitimate feeling in these circumstances and needs to be acknowledged. However it was felt by most people consulted that that anger and/or blame should not be expressed in a memorial." The paper quoted the views of one community member who stated, "There are other processes for that, the memorial should be about remembrance and gratitude."[47]

The consultation process defined the aspects of the disaster considered appropriate or desirable to address within a commemorative space. Anger was defined as outside the scope of "remembrance." Instead, the paper defined the purpose of the memorial in three parts: First, "commemorative" in that it would acknowledge lives lost, the scale of impact and "honour the courage of those who risked their lives to help others and to express gratitude"; second, "restorative" in that it would "celebrate the resilience and spirit of our community and the beginnings of regeneration of a much loved environment" and, third, "educative," in that it would "ensure the lessons of the 2003 experience are learned and encourage better bushfire preparedness at all levels in our community."[48]

Between publication of the consultation paper and of the committee's final report, however, the "educative" role of the memorial was removed. The final report argued, "The recommendation to take out the educational role was not because bushfire education was seen as unimportant. On the contrary, it was because of its high importance and the need for specialist input."[49] The views of community members on this topic were quoted, including "The educative role is too large and too complex to be held in the memorial"; "Just by having a memorial will be enough to get people thinking"; and "Separate education from the memorial as education is not peaceful."

[47] RPR Consulting, "A Bushfire Memorial for the ACT: Community Consultation Discussion Paper" (Canberra, ACT: artsACT, Department of Urban Services, 2004), 6.

[48] RPR Consulting, "A Bushfire Memorial for the ACT": 9.

[49] RPR Consulting, "A Bushfire Memorial for the ACT: Community Consultation: Final Report" (Canberra, Vic: artsACT, Department of Urban Services, 2004), 9.

The report also saw community anger and education as interwoven stating, "Strong feelings about bushfire education can become merged with anger about a perceived lack of early warning about the threat posed by the January bushfires." The report reiterated that the memorial was not considered an appropriate place for the expression of anger. A community member was quoted in the report as stating: "Let's not go there. That is not what the memorial should be about. It's about commemorating those who died and celebrating the way the community rallied together and survived. Let's leave it at that."[50]

Removing anger from the memorial removed a central aspect of the disaster from the site at which it was to be commemorated, in line with historian Peg Fraser's observation that, "It is part of the process of commemoration to edit the past."[51] The removal of anger also removed the possibility of effective bushfire education. Education may not be peaceful, but it is necessary. Anger at perceived flaws in policy might disrupt opportunities for quiet reflection, but might also produce changes in community behaviours and in government policies with future benefits. As configured in the final report, the memorial was to be developed as place to both remember and forget.

2006: The Third Anniversary

A year later, the memorial was complete and ready to be officially opened. Designed by artists Tess Horwitz, Tony Steele and Martyn Jolly, the site begins with entrance walls constructed from salvaged bricks donated by the community and inscribed with messages. Four bricks are inscribed with the names of the people who died. Others are decorated with messages to lost pets, statements of gratitude and celebrations of resilience. A path through the entrance leads to an open area containing five curved steel-girders, designed to suggest trees bent by the force of strong winds in the firestorm. An avenue then leads alongside a small artificial creek which ultimately feeds into a large reflection pond. Positioned around the pond

[50] Ibid., 9.

[51] Peg Fraser, *Black Saturday: Not the End of the Story* (Carlton, Vic: Monash University Publishing, 2019), 78.

are five pillars, each of which contains photographs donated by survivors and displayed behind glass. It is an attractive, immersive space surrounded by trees and shrubs, which encourages the kinds of quiet reflection that the community consultation process had adopted as the space's main objective. At the opening ceremony one survivor, Robyn Middleweek, told the *Chronicle*, "I think it is wonderful. It's a calm place."[52]

Tensions had yet to calm, however, between some (although certainly not all) survivors and the ACT Government. Issues of continued, heated debate included government interventions in the coronial inquest; the level of prescribed burning in place before the fires; the amount of financial support provided to survivors; the speed of developmental approval processes and unresolved anger over emergency management decisions. Across several Canberra suburbs, hundreds of residential blocks sat empty. Only half of the destroyed houses had been rebuilt. Memories of the disaster were etched into the suburban landscape and into the minds of recovering individuals. A memorial had been constructed, and yet, for many, the disaster had yet to end.

It is possible to read the government funded memorial site as an attempt at foreclosing ongoing controversies. As Simpson and de Alwis argue, a memorial reflects "the concerns of those who put it there, and those who control the public activities that may surround it."[53] A memorial might contain an implicit message that the political response to a disaster is now complete and it is time to look to the future. Those who continue to publicly display their anger are implicitly criticised for their inability to "move on." In the days prior to the ceremony, an ACT Government spokesperson asked employers to allow staff time to attend the event and stated, "This is expected to be the last 'official' recognition of the bushfire anniversary."[54]

It is also true, however, that the development of the memorial as a site for peaceful reflection, not anger or education, was in line with views many survivors had expressed during the disaster's first three years. Asked about a possible memorial in 2004, Alan Latta had stated, "It would be nice to sit in a bit of peace, even if there's nobody there you can sit and

[52] "Bushfire memorial unveiled," *The Chronicle*, 24 January 2006, 5.

[53] Edward Simpson and Malathi de Alwis, "Remembering natural disaster: politics and culture of memorials in Gujarat and Sri Lanka," *Anthropology Today* 24, no. 4 (2008): 7.

[54] "Fitting tribute to all creatures great and small," *Canberra Times*, 13 January 2006, 4.

reminisce and think about things and have a quiet time. Because that's what it would mean to me, a quiet time."[55] Also in 2004, Emma Walter had stated, "Whatever the memorial will reflect will hopefully be something that reflects that we got through it sort of thing, rather than reminiscence about the bad, maybe the sort of more focus on the positives that we survived it, we got through it."[56]

For survivors, the disaster was unforgettable. Their goal was to find ways of remembering which were peaceful and which did not revive trauma. For some, this meant a refusal to take part in official commemorative events. For others, it meant participation in ways through which they could make their own meanings and draw their own value. As the Farrells stood with their protest sign on the morning of the third anniversary, their objection was not to the site but to the presence of the Chief Minister. They wanted the target of their anger to be removed, so that they could reflect in peace.

This memorial was developed as a space in which to be reminded that recovery was possible, not that trauma was ongoing or that future bushfires were likely. In her speech at the opening ceremony, Jane Smyth stated, "We were all involved in some way in a terrifying and shocking event and we need to sort our thoughts and memories of what did happen. … And remembering that day and sorting its events in our minds is part of moving on."[57] Three years from the fires, "moving on" was still a process in which survivors were participating. The disaster had not been left behind, but commemorative events and sites were being utilised as mechanisms through which to safely remember.

Conclusion

According to Andrew Salvador Matthews, "Fire is a powerful symbol of chaos, often marking the destruction of order in both the social and the natural worlds."[58] Disaster commemoration is a way of re-establishing

[55] Latta, Interview with Mary Hutchison.

[56] Walter, Interview with Mary Hutchison.

[57] "The blackest of days, 3 years ago, remembered," *Canberra Times*, 19 January 2006, 1.

[58] Andrew Salvador Mathews, "Suppressing Fire and Memory: Environmental Degradation and Political Restoration in the Sierra Juárez of Oaxaca, 1887–2001," *Environmental History* 8, no. 1 (2003): 77.

order. It is a collective act of meaning-making, through which the present is constituted as a well-ordered moment distinct from the disaster's disarray. While anniversary events stressed narratives of strength and resilience, the bushfire memorial came to be defined as a place of quiet, peaceful and ordered reflection. This is an understandable response from a community so catastrophically disrupted by disaster. In the three years examined in this chapter, commemoration was occurring in a landscape defined by what the fire had removed and among traumatised people still struggling to recover. In this context, there was a clear need to hear stories of heroism and resilience and to find a still place for quiet expressions of gratitude.

There is an inherent tension in disaster commemoration between conflicting desires to remember and to forget. A memorial is an attempt both to fix memories of the disaster in a constant present—always remembered, never forgotten—and simultaneously to "move on" into a new future, leaving the disaster in the past. Simpson and de Arwis argue that a memorial "translates the unthinkable into the thinkable."[59] Survivors will never forget the disaster, but they sought ways of remembering that did not revive trauma.

As a consequence, however, it is possible that important truths became detached from the commemorative process. Survivors may not need reminding of the devastating impacts of fires, but the broader community does. Over time, as the population of Canberra changes and a wave of new residents move into rebuilt suburbs, the memory of the fires will fade. Anniversary events, if held at all, will draw smaller crowds. The memorial site, sitting peacefully out of view, will become one of the large number of heritage attractions sprinkled throughout the capital. The kinds of active remembering essential to encouraging community members to prepare for future fires and to understand past policy errors do not exist within the memorial as it stands and will require specific action to prevent another disaster.

[59] Simpson and de Arwis, "Remembering natural disaster": 7.

Part II

Human Understandings of Disaster

"Shaken but Not Stirred": The Aftermath of Urban Disasters

Margaret Cook

After Newcastle's 1989 earthquake, the local daily newspaper, *The Newcastle Herald* ran the banner headline "Shaken but not stirred: the spirit of Newcastle lives on despite disaster".[1] A familiar nod to the unflappable film character of 007 fame, James Bond, the headline highlights the dimensions of disasters explored in this chapter. The word "shaken" is an admission that the earthquake had challenged Newcastle residents' notion of place and sense of security. The phrase "not stirred" was intended by editor, John Lewis, as a reassurance that the city would recover.

Lewis's article claimed Newcastle residents, who adopt the moniker Novocastrians, "don't object to change—as long as it doesn't destroy the human scale of the place". But on 28 December 1989 at 10.27 a.m., the

[1] "Shaken but not Stirred: The spirit of Newcastle lives on despite the disaster", *Newcastle Herald*, 1 January 1990, 3.

M. Cook (✉)
School of Social Sciences, University of the Sunshine Coast,
Sippy Downs, Australia
e-mail: mcook1@usc.edu.au

S. McKinnon, M. Cook (eds.), *Disasters in Australia and New Zealand*,
https://doi.org/10.1007/978-981-15-4382-1_5

"unthinkable" happened. An earthquake fundamentally altered the city, taking 13 lives, injuring 162 people and damaging 10,000 buildings (112 homes and 192 commercial buildings partially or completely destroyed) (Fig. 1).[2] In a narrative common to disaster reporting,[3] Lewis extolled the virtues and courage of the rescuers, drawing on Newcastle's "resilience and vitality". He wrote, "Our city will be changed but its spirit remains intact".[4] The article reflected a deep-seated human need to reassert control over the environment.

Brisbane reflected this trend after the 2011 floods with the *Courier-Mail* headline, "Defiance. As the flood of 2011 recedes, the fightback begins".[5] On 13 January floodwaters rose to 4.46 metres at the Brisbane Port Office flood gauge, taking one life, inundating 94 suburbs and damaging 14,100 Brisbane properties.[6] Many Brisbane residents had believed that the Wivenhoe Dam, opened in 1985 had flood-proofed their city, but this belief was shattered. The population was clearly stirred, surprised that nature had dispelled human notions of environmental security. As in Newcastle, the media recounted tales of heroism and resilience shown by Brisbane citizens, adding to politicians' reassurances that the city would recover. Flood stories were framed by the community as a celebration of resilience and heroics, a statement of what is required to be a Queenslander.[7] The *Courier-Mail* called on the state's residents to stimulate recovery with the editorial "Resilience to grow from depths of this great flood",

---

[2] "The Newcastle Experience- Earthquake Renewal: 1990–1993," *Report of the Newcastle Region Renewal Co-ordination Unit*, 4, 1995, 4; Linda Doherty, Mike Scanlan and Roderick Quinn, "Down below, many of the 10,000 casualties," *Newcastle Herald*, 4 January 1990, 1; Finding Newcastle Earthquake Disaster of the 28th December 1989, Findings delivered 18th July, 1990, 24–25, http://www.coroners.justice.nsw.gov.au/Documents/newcastle%20earthquake%201989%20-%20finding%20and%20recommendations.pdf.

[3] Mervi Pantti, Karen Wahl-Jorgensen and Simon Cottle, *Disasters and the Media* (New York: P. Lang, 2012), 85–86; Brian Miles and Stephanie Morse, "The role of news media in natural risk and recovery," *Ecological Economics*, 63, (2007): 366.

[4] "Shaken but not Stirred", 3.

[5] "Defiance. As the flood of 2011 recedes, the fightback begins," *The Courier-Mail*, 14 January 2011, 1.

[6] Queensland Floods Commission of Inquiry (QFCI), *Interim Report* (Brisbane: Queensland Floods Commission of Inquiry, 2011), 27.

[7] Scott McKinnon, "Remembering and Forgetting 1974: The 2011 Brisbane Floods and Memories of an Earlier Disaster," *Geographical Research*, 57, No. 2 (2019): 204–214.

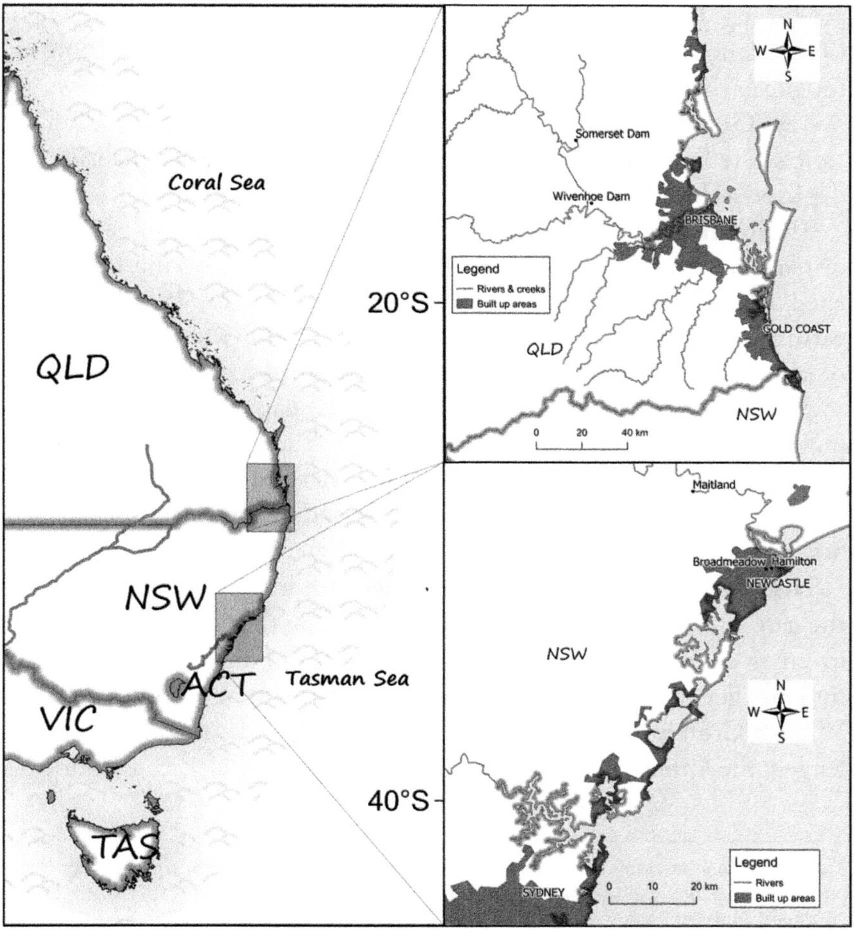

Fig. 1 Map of locations mentioned in this chapter

publishing tales of "generosity and stoicism".[8] At the flood's height, Queensland Premier Anna Bligh tapped into this created self-identity of tough survivors as she rallied locals: "I want us to remember who we are. We are Queenslanders".[9] Although traits of courage, determination and stoicism reflect disaster responses around the globe, these qualities were depicted as locally unique.

The human responses in both cities reflected deeply embedded notions of place, environmental security, and control and these disasters threatened these imagined realities. This chapter uses the 1989 Newcastle earthquake and the 2011 Brisbane flood to explore human conceptions of place and its nexus with nature. As Peter Read maintains, "place" is a social construct, created by an individual and shaped by societal forces of power and ideology.[10] A locale is imbued with meaning, embedded with emotional and psychological values that relate to "identity, belonging, security, self-esteem, self-efficiency and solace".[11] Cities are socially produced, shaped by "physical and socio-economic flows".[12]

Historically, cities have been imagined and described as separate from the natural world, a human creation imposed on the land, distanced from its environmental context.[13] A city provided the ultimate expression of human agency and mastery of nature, creating a new landscape. Led by William Cronon's definitive work, *Nature's Metropolis,* scholars have challenged the urban/nature divide, arguing that the city is a hybrid—a

[8] "Resilience to grow from depths of this great flood," *Courier-Mail,* 12 January 2011, 26; "Sirens puncture the silence of a slow-moving disaster," *Courier-Mail,* 13 January 2011, 79; Trent Dalton, "Kindness comes to muddy street," *Courier-Mail,* 15–16 January 2011, 6. Other examples include "Coping with crisis," *Sunday Mail,* 2 January 2011, 42.

[9] Anna Bligh speech, 13 January 2011, http://statements.qld.gov.au/Statement/Id/73282.

[10] Peter Read, *Returning to Nothing: The Meaning of Lost Places* (Cambridge, England: Cambridge University Press, 1996), 2.

[11] Neville R. Ellis and Glenn A. Albrecht, "Climate change threats to family farmers' sense of place and mental wellbeing: A case study from the Western Australian Wheatbelt," *Social Science and Medicine,* 175 (2017): 161.

[12] Lisa Benton-Short and John Rennie Short, *Cities and Nature,* 2nd edition (London: Routledge, 2013), 2.

[13] Joel A Tarr, "The City as an Artefact of Technology and the Environment," in Martin Reuss and Stephen H Cutcliffe, eds., *The Illusory Boundary: Environment and Technology in History* (University of Virginia Press, 2010), 145; Martin V. Melosi, "Humans, Cities and Nature: How do Cities Fit in the Material World?" *Journal of Urban History,* 36, no. 1: 3–21.

manufactured human space and also an integral part of nature.[14] As Lisa Benton-Short and John Rennie Short state: "there are no cities independent of nature".[15]

There is inherent reciprocity between cities and nature, argues Joel Tarr.[16] Cities alter (or in more declensionist language, exploit and destroy) nature but in turn the natural forces of climate and seismic shifts impact on urban environments. Floods and earthquakes reveal that nature is ever-present in the city. Yet for many metropolitan dwellers, earthquakes and floods have no place in Australia's geological and hydrological imaginaries, nor in their cities.

Despite the *Newcastle Herald* editor's claims to the contrary, Newcastle and Brisbane residents were deeply stirred. Their sense of place and safety (or invincibility) against nature was challenged. Disasters offer a significant moment of societal disruption, an opportunity to change path dependency on policies and practices that have created the hazard and increased risk within the city. But, economist Paul David argues, the opportunity for change exists only briefly, before the status quo is reinforced.[17] In Newcastle and Brisbane the opportunity was lost as decision-makers focussed on rebuilding, a political imperative that Tom Griffiths shows "constrains cultural adaptation".[18] Both city's residents were "not stirred" as they avoided difficult questions about human agency in natural disasters. In Newcastle, community anguish shifted to arguments about salvage and reconstruction of earthquake-damaged buildings. In

[14] William Cronon, *Nature's Metropolis: Chicago and the Great West* (New York, W. W. Norton, 1991). See, for example, Martin Melosi, "The Place of the City in Environmental History," *Environmental History Review*, 17, no. 1 (1993): 1–23; Christine Meisner Rosen and Joel. A Tarr, "The Environment and the City," *Journal of Urban History*, May (1994): 299–334; Jordan Bauer and Martin V. Melosi, "Cities and the Environment," in *A Companion to Global Environmental History*, eds. John Robert McNeill and Erin Stewart Mauldin (West Sussex, UK: John Wiley and Sons, 2012), 360–377.

[15] Lisa Benton-Short and John R. Short, *Cities and Nature* (London: Routledge, 2013), 2.

[16] Joel A. Tarr, "Urban History and Environmental History in the United States: Complementary and Overlapping Fields," in *Environmental Problems in European Cities in the 19th and 20th Century*, ed. Christoph Bernhardt (Munster: Waxman, 2001), 25–39.

[17] Paul David, "Historical Economics in the Longrun: Some Implications of Path-Dependence," in *Historical Analysis in Economics*, ed. Graeme Donald Snooks (London: Routledge: 1993), 38–39.

[18] Tom Griffiths, "The Language of Catastrophe: Forgetting, Blaming and Bursting into Colour," *Griffith Review*, 35 (2012): 52.

Brisbane, wrath was directed at presumed dam mismanagement. In both cases these responses reflected the desire to reassert a sense of place, security and human control over nature.

A Sense of Place

Measuring 5.6 on the Richter scale, with its epicentre 15 km south west of Newcastle, this was the biggest earthquake in southern-eastern Australia in the twentieth century, felt throughout a quarter of New South Wales (NSW), a state more than twice the size of Germany.[19] Earthquakes are not unknown in Australia. The Awabakal people of Newcastle remember past occurrences recounted in their Dreaming stories.[20] Since British colonisation in 1788 Australia has experienced 20 magnitude 6 earthquakes, with one as large as Newcastle's occurring every two years (up to 2010), the most substantial in 1837, 1841, 1842, 1868 (5.3 magnitude), 1925 (5.0) and the largest, 1989 (5.6).[21] Seismologists warned in 1988 that the Hunter region, where Newcastle resides, was a high earthquake risk area with 55 earthquakes felt within 100 km of the city since 1950. As most occurred in small, uninhabited areas, these were inconsequential.[22] The 1989 earthquake was different. Newcastle had become a major urban centre (the second largest city in NSW) and this "killer quake" claimed

---

[19] Finding Newcastle Earthquake Disaster of the 28th December 1989; Terry Smith, "Quake proves experts correct," *Newcastle Herald*, 29 December 1989, 4; Lenore Taylor, "Quarter of State felt earth move," *The Australian*, 29 December 1989, 4.

[20] One legend describes the creation of the passage between Points Nepean and Lonsdale in NSW. See Kevin McCue, "Australia: Historical Earthquake Studies," *Annals of Geophysics*, 47, no. 2/3 (2004): 387. The other is about Nobbys Head (Whibayganba) in Newcastle. See NSW Government Education Standards Authority, *The Kangaroo that Lives Inside Nobbys*, 2008, https://ab-ed.nesa.nsw.edu.au/go/7-10/science/units/story-2/units/introducing-our-changing-earth-/activities-and-worksheets/the-kangaroo-that-lives-inside-nobbys.

[21] Cynthia Hunter, *The Earth was Raised up in Waves like the Sea…: Earthquake Tremors Felt in the Hunter Valley since White Settlements* (Newcastle: Hunter House Publications 1991), 1; Kevin McCue, "Assessing Earthquake Hazard and Risk in Australia," *Australian Planner*, 47, no. 1 (2010): 52.

[22] Smith, "Quake proves experts correct", 4; Finding Newcastle Earthquake Disaster of the 28th December 1989, 5.

lives.[23] Anthropogenic factors—building on alluvial soil and infill areas, the central business district (CBD) having been constructed on a former course of the Hunter River—substantially increased the impact.[24] Reclaimed land had increased the city's vulnerability.

Despite this seismological history, when an earthquake struck Newcastle, the general public was stunned, disbelieving; earthquakes were not associated with Newcastle. Described by the *Newcastle Herald* as "unprecedented", witnesses struggled to comprehend the event.[25] Local journalist Clare Morgan considered a bomb or gas explosion, an earthquake "doesn't enter your head in Newcastle".[26] In this industrial city some considered an explosion at BHP or mine subsidence, others a plane crash or tornado.[27] Many residents only accepted an earthquake occurrence after witnessing widespread damage or heard media confirmation.[28] Greg Laurie expressed the thoughts of many—"this was something that happens in another country".[29] Standing outside his damaged terrace house, Andrew Turnbull concurred, "This doesn't happen in Newcastle—it's not San Francisco".[30]

In 29 seconds, the earthquake changed Newcastle, physically and psychologically. Nine people died in the collapsed Workers Club, three died when buildings fell on them in the suburb of Hamilton (5 km west) and

[23] The phrase "killer quake" was employed as a banner in *The Age*, 29 December 1989 and "Death toll mounts from killer quake," *The Courier-Mail*, 29 December 1989, 2.

[24] Finding Newcastle Earthquake Disaster of the 28th December, 1989, 1; R. E. Melchers, "The Newcastle Earthquake—Recommendations for the future," in BSFA Seminar, *An Earthquake in Sydney- What have we learnt from Newcastle?* 20 June 1990, 23.

[25] "A Newcastle Tragedy," *Newcastle Herald*, 29 December 1989, 4.

[26] Clare Morgan, "A bumpy road, a set of broken traffic lights, then a city in shock," *Newcastle Herald*, 29 December 1989, 5.

[27] "Two crushed to death in Hamilton 'war zone'," *Newcastle Herald*, 29 December 1989, 7; "I thought it was a tornado, I was so shaken up," *Newcastle Herald*, 29 December 1989, 5.

[28] "Street filled with dazed and frightened people," *Newcastle Herald*, 29 December 1989, 9; Peter Barrack, oral history interview, Tape 2, CY MLOH 559, Mitchell Library.

[29] "A Newcastle Tragedy," *Newcastle Herald*, 29 December 1989, 4; Greg Laurie, oral history interview, Tape 10, 2 January 1991, CY MLOH 559, Mitchell Library. See also Bob Beale, "An eerie silence envelops city centre," *Sydney Morning Herald*, 29 December 1989, 2; "Residents tell of lucky escapes and dreams shattered," *Newcastle Herald*, 29 December 1989, 5; Jean Thomas to Dorothy Anderson, 17 January 1989, 4. MLMSS 5227 Mitchell Library.

[30] Adam Walters, "This doesn't happen in Newcastle —it's not San Francisco," *Courier-Mail*, 29 December 1989, 2.

one died of a shock-induced heart attack.[31] At the Workers Club, the western wall had collapsed, the floors of the auditorium, poker machine level and car park fell into the basement, pinning patrons and staff under rubble for hours. A huge slab of concrete sheared free and lay on a steep incline into the basement below, while "thick steel reinforcing rods dangled like spaghetti".[32] Extensive damage in Beaumont Street, Hamilton, required 40 buildings to be secured, with 300 incurring parapet or suspended awning failure. Chimneys, gables, and church spires and towers proved vulnerable, as the top of the war memorial lay unceremoniously on the ground, the Obelisk on The Hill left in danger of collapse.[33] The Centenary Theatre at Broadmeadow resembled a "bomb site", while the Junction Motor Hotel balanced precariously on its pylons. An estimated 50 per cent of buildings in the CBD and 75 per cent of buildings in Hamilton would require demolition or considerable repair.[34] Many lost houses they had lived in for decades, 150 people faced homelessness, a problem compounded as destroyed workplaces brought unemployment, with 4000 people receiving financial assistance.[35] Dust shrouded Newcastle. Shock set in, the realisation for many that "the life they knew and loved had gone".[36] According to *The Age*, the "overriding emotions were sadness, at the rising toll of life and landmarks, and disbelief … an earthquake was never a possibility".[37] Earthquakes became entwined in Newcastle's sense of place.

In promotional material, Brisbane calls itself "The River City", in its imagined geography the Brisbane River is regarded as a benign aesthetic feature and source of profitable riverside land. But in this sub-tropical

---

[31] Finding Newcastle Earthquake Disaster of the 28th December, 1989, 24–25.

[32] "One glance shattered illusion over devastation," *Newcastle Herald*, 29 December 1989, 3.

[33] "Century Theatre left crumpled resembling bomb site after shock," *Newcastle Herald*, 29 December 1989, 5; "Junction motel faces $2m damage bill," *Newcastle Herald*, 29 December 1989, 6; "Newcastle declared disaster area after deadly quake", *Newcastle Herald*, 29 December 1989, 2.

[34] Sybil Nolan, "Two bodies make quake toll 12," *The Australian*, 1 January 1990, 1.

[35] Cathy Johnson and AAP, "More than 150 people left without a home," *Sydney Morning Herald*, 30 December 1989, 5.

[36] Denis Butler, "Luck turns to shock as Thursday's nightmare strikes with brutal reality," *Newcastle Herald*, 1 January 1990, 5.

[37] Andrew Kovacs, "Newcastle tries to come to terms with the impossible," *The Age*, 1 January 1990, 7.

climate, the river floods, the most disastrous recorded in 1841, 1893 (twice) and 1974, with smaller floods in between. In the 24 hours preceding 9 am on 10 January, 300 mm of rain fell in the Brisbane catchment, a further 700 mm falling in the next 24 hours.[38] The Brisbane River system was in flood as Wivenhoe Dam received 2.65 million megalitres forcing substantial releases. The floodwaters irreparably altered the city's fabric. In total 14,100 properties were affected (1203 houses, 1879 businesses partly inundated and 557 completely submerged).[39] Suburbs lay submerged as roof tops, street signs and telephone poles peaked out from the floodwaters, phone lines and electricity cut. Roads and railway corridors were reduced to canals, the iconic Riverwalk smashed by the debris adrift in the river. Stinking mud-coated buildings and possessions, lifetimes of memories, formed piles of rubbish on the footpaths. As Premier Anna Bligh acknowledged, there have been "lives lost, homes destroyed, and livelihoods washed away in the cruellest and most indiscriminate way". She recognised community sentiment: "the past few weeks have left us shocked and shaken".[40] At least in the immediate future, floods would gain recognition as an environmental characteristic of a city where its residents preferred to ignore them, but a dominant rhetoric would quickly help re-establish an imagined landscape where the environment was controlled.

Security Threatened

Floods defied Brisbane residents' perception of flood immunity. Brisbane had last endured a devastating flood in 1974 and for 37 years many people thought floods would never happen again, a belief underpinned by a reliance on the city's dams. Somerset Dam, completed in 1958, had reduced the 1974 flood height, but the 1984 completion of Wivenhoe Dam, with over three times the flood storage, would, many believed,

[38] Australian Government Bureau of Meteorology, *Report to Queensland Floods Commission of Inquiry*, March 2011, 13–17, 54.

[39] Queensland Floods Commission of Inquiry, *Interim Report*, 27.

[40] Anna Bligh, "Message of support from the Premier," *Courier-Mail*, 19 January 2011, 36.

provide the desired flood prevention. Although engineers repeatedly emphasised dams can only mitigate floods, their finite storage capacity requiring water releases, residents clung to the myth that the "mighty Wivenhoe Dam", Brisbane's "great protector", would keep Brisbane flood-free.[41]

After the 2011 flood residents expressed surprise that the river had reneged on its manufactured role as a benign aesthetic and economic asset. *Courier-Mail* editor Christine Middap revealed her sense of the river's treachery: "That big, brown powerful river. Every time I look at the Brisbane River I will remember what she did to us; what she is capable of".[42] Mike Bruce, at the *Sunday Mail*, suggested that humans should file for divorce against the Brisbane River. "It would", he wrote, "be only human to feel a sense of betrayal or contempt after the way our river, our beloved river, rose up and struck out so violently at the very people who took it into their hearts". He added, "importantly, post-Wivenhoe, we were confident it would never again turn on us with the vengeance it did in 1974".[43] But this was no act of betrayal. Floods are an essential process in the river's cycle, the floodplain an integral part of the river.

Nature had shattered the pervasive myth of flood immunity and revealed the foolishness and arrogance in believing nature could be tamed. Journalist Robert MacDonald expressed the common view that Brisbane, post-Wivenhoe Dam, "would no longer be captive to its environment" but the floods had shown Brisbane's floodplain dwellers that they did not, as the editor of Australia's national newspaper, the *Australian*, explained, "have the control of their own lives [as] they assumed".[44] The flood had exposed the city's vulnerability.

---

[41] "State of Disaster," 'The 2010–2011 Floods Crisis', APN Special Commemorative Publication, February 2011, p. 35; Anna Caldwell, Renee Viellaris and Sarah Vogler, "Devastation on a scale never seen before," *Courier-Mail*, 13 January 2011, 2; Zane Jackson, "City braces for water to peak," *Queensland Times*, 12 January, 2011, 14; QWeekend, "Facing Armageddon," 20 January, 2011, https://www.youtube.com/watch?v=ZdWhYLMDRUQ.

[42] Christine Middap, *QWeekend editorial*, "The Flood 2011," 3.

[43] Mike Bruce, "Love affair with the Brisbane River," *The Sunday Mail*, 23 January 2011, 30. See also Shane Rodgers, APN Special, 3.

[44] Robert MacDonald, "Bureaucrats, miners, even God in firing line as blame game begins," *Courier-Mail*, 19 January 2011, 27; "Disaster unites city and country as never before," *The Australian*, 13 January 2011, 15.

In the immediate aftermath of the earthquake, Peter Barrack, President of the Newcastle Workers Club, recalls people walking around "stunned", trying to "come to grips" with what and why it had happened.[45] As resident Jean Thomas wrote to her friend Dorothy Anderson, "It was as if each and every one of us were experiencing a mass nightmare, too weird and horrific to contemplate. We waited to come out of it, but that didn't happen".[46] The dream of environmental security had soured.

Local journalist Kim Britton noted, the "aftershock of fear and dislocation" lingers. Novacastrians were struck by "a feeling of frailty, vulnerability", a sense of "helplessness in the face of the quake has left a legacy of heightened mortality". [47] Resident, Jean Thomas, told her friend, "the worst legacy—however is a sense of helplessness, insecurity and fear. Everything makes you jumpy, every day is the fear that it could happen again".[48] *Courier-Mail* journalist Juanita Phillips wrote, "along with buildings and human lives, something else was shattered by the Newcastle earthquake: national complacency".[49] The editor of Sydney's daily newspaper, the *Sydney Morning Herald,* wrote what no one wanted to read. The earthquake provided a "clear reminder that Australia is not immune" from earthquakes. It "may happen again".[50] As structural engineers Professor Rob Melchers and Associate Professor Adrian Page from Newcastle University stated, "the illusion of perfect safety has surely been shattered".[51]

---

[45] Peter Barrack, oral history interview.

[46] Jean Thomas to Dorothy Anderson, 17 January 1989, 1.

[47] Kim Britton, "After the clean-up, the shared aftershock of disruption," *Newcastle Herald,* 1 January 1990, 2.

[48] Jean Thomas to Dorothy Anderson, 17 January 1989, 1.

[49] Juanita Phillips, "No longer a complacent Australia," *The Courier-Mail,* 29 December 1989, 9.

[50] Editorial, "Lessons from an earthquake," *Sydney Morning Herald,* 30 December 1989, 14.

[51] Paul Chamberlin, "Poor construction held to blame," *Sydney Morning Herald,* 6 January 1990, 5.

Re-establishing Control

Recovery is the modus operandi after any disaster, re-establishing some degree of normality as soon as possible. In his work on the 2005 Hurricane Katrina in New Orleans, Steve Kroll-Smith has shown how authorities in the United States of America worked to assert control and re-establish the social structure and market forces.[52] Common to all disasters the State's power elite—politicians, property and business owners—have a vested interest in a quick recovery.[53] Such was the case in 1974 after Cyclone Tracy hit Darwin, where, as Peter Reid shows, the Darwin Reconstruction Commission's task to rebuild prompted a battle with the local community.[54] Similarly, after the 2010 and 2011 earthquakes in Christchurch, New Zealand, Bronwyn Hayward notes that under the guise of resilience, authorities made decisions quickly, success measured by the speed by which the city attained its "new normal" or certainty. She argues that in the pursuit of re-establishing order, democracy can be the casualty with expert command-and-control decisions made with limited local scrutiny or community leadership.[55]

As seismologists reassessed the country's risk of future earthquakes,[56] Novocastrians endeavoured to restore their city. The *Newcastle Herald* editor described the situation:

> Newcastle will face a long and hard slog as it tackles the task of healing its wounds. The bereaved will need to be comforted, the injured cared for, the homeless housed.

The editor added, "the earthquake was a brutal blow, individually and collectively, but Newcastle's powerful community bonds will cope with

[52] Steve Kroll Smith, *Recovering Inequality: Hurricane Katrina, the San Francisco Earthquake of 1906, and the Aftermath of Disaster* (Austin: University of Texas Press, 2018).

[53] Margaret Cook, "'It will never happen again': The Myth of Flood Immunity in Brisbane," *Journal Australian Studies*, 42, no. 3, (2018), 331.

[54] Reid, *Returning to Nothing.*, 158–165.

[55] Bronwyn Hayward, "Rethinking Resilience: Reflections in the Earthquakes in Christchurch, New Zealand, 2010 and 2011," *Ecology and Society*, 18, no. 4, 37.

[56] Smith, "Quake proves experts correct," 4.

its effects".[57] Newspapers called on parochial pride to foster resilience and recovery.

Alongside deaths, injury and damage to homes and businesses, Novocastrians had to endure further loss, the demolition of damaged heritage buildings. Authorities concerned about aftershocks and public safety were keen to re-establish control, with Day 12 (8 January) the target for re-opening the city.[58] With the Council in recess, emergency measures were implemented by the State Emergency Service, fire brigade, police, ambulance and army.[59] These people determined the fate of damaged buildings, not the elected Council, as noted in Christchurch.[60] Authorities acted swiftly to reassert a sense of control over nature.

Writing about Christchurch's 2011 earthquake, historian Katie Pickles identified that people, faced with loss and grief, "sought comfort in the past", manifesting in debates on how to rebuild.[61] Abrupt demolition decisions in Newcastle evoked the same response. The need to remove bodies brought immediate partial demolition of the Workers Club, soon followed by the razing of the newly restored Kent Hotel. Earthquake damage to the city's heritage was hard felt, with Jean Thomas writing: "To see our wonderful old churches, schools, libraries, our heritage destroyed, was heartbreaking". While she had endured bushfires, this disaster seemed worse. "It won't blow over, the buildings—unlike the bush, won't grow back, the overwhelming sense of loss will stay with us, the whole thing is too huge to handle".[62] For Jean, these buildings shaped Newcastle's identity.

The *Newcastle Herald* noted the additional distress demolition caused. "Newcastle will mourn the people who have died, its loss of earthquake innocence and the familiar buildings that are being extracted like

[57] "A Newcastle Tragedy," *Newcastle Herald*, 29 December 1989, 1.

[58] Harold Stuart, "The Newcastle Earthquake: Local Government Response," *The Macedon Digest* 7, No. 4 (1992), 18.

[59] Margaret Henry, "The battle for Newcastle: heritage and the earthquake," *Australian Historical Studies*, 24, no. 96 (1991), 103. This article provides a detailed account of the heritage battle.

[60] Henry, "The battle for Newcastle," 108.

[61] Katie Pickles, *Christchurch Ruptures* (Wellington: Bridget Williams Books Limited, 2016), 128–129.

[62] Jean Thomas to Dorothy Anderson, 17 January 1989, 1–2.

crumbling teeth from a once-complete smile".[63] The Novacastrians' sense of place was being destroyed by the need for control. Fears abounded about possible demolition of churches, shops and houses; places identified by the National Trust as having heritage significance. But as Pickles found in Christchurch, "not all rubble" was equal, a higher value assigned to statues and heritage buildings, reminders of New Zealand's colonial past.[64] In Newcastle, where many residents clung to these icons of history, officials informed the media that "much of the city would have to be demolished and rebuilt".[65] Newspapers expressed community fears that "tragically, the very heritage of Newcastle is in danger of disappearing".[66] A *Newcastle Herald* article headlined "Half Beaumont St facing demolition: Hamilton valuer" further exacerbated concern.[67] On assessing the damage, CSIRO building engineer expert, Dr George Walker, declared, "many of these buildings were close to the end of their economic life". He added, "the best policy on economic grounds is demolition and replacement with more modern structures better fitted for modern living". He warned that the cost of repair may out way the value of the building.[68] But as heritage practitioners know, the value of heritage is often cultural, not financial.

Heritage advocates rallied, raising concerns about a "precipitate rush" to demolish when repair would suffice, with poor decisions made by "well meaning, but ignorant" people without heritage expertise.[69] Residents formed the Citizen Earthquake Action Group with the aim of restoring the "city back to its original fabric of life before the earthquake".[70]

---

[63] "Shaken but not Stirred," 3.

[64] Pickles, *Christchurch Ruptures*, 16–17.

[65] Scott Symonds, "Quake toll could reach 15," *Sunday Age*, 31 December 1989, 1.

[66] Brian Woodley, "Engineer has yet to inspect a building he can pronounce safe," *The Age*, 30 December 1989, 6.

[67] "Half Beaumont St facing demolition: Hamilton valuer," *Newcastle Herald*, 30 December 1989, 5.

[68] George Walker, *Interim report on Newcastle Earthquake* (North Ryde, NSW: CSIRO Australia, 1990), 12.

[69] "Building codes 'need review'," *Newcastle Herald*, 30 December 1989, 9; J. W. Jordan, J. N. Ludlow and E. G. Trueman, "The Newcastle Earthquake and Heritage Structures," Fifth National Conference of Engineering Heritage 1990, Perth, 3–5 December 1990, 108.

[70] "Heritage watchdog group moves to protect buildings," *Newcastle Herald*, 4 January 1990, 10.

The group lobbied State and Federal Governments to advise caution and provide external experts, otherwise, conservationist Chris Richards warned, Newcastle would "end up with a faceless miners' city without its heritage".[71] Their actions reflected a fundamental desire to preserve an imagined local identity.

More than 150 people from conservation groups gathered on 31 December 1989 to protest. Participant Margaret Henry and member of the Northern Parks and Playground Movement reported residents "felt they were being bullied out of their heritage and their homes". She added,

> The future of our city is in the hands of people who have little knowledge or understanding of the need to preserve its unique character. We care about our homes and we care about our heritage. We feel powerless. The earthquake has been traumatic enough with the terrible loss of life. This should not be made worse by the destruction of our homes and our heritage.[72]

The conflict grew as heritage advocates accused Council and property owners of profiteering from the earthquake, taking the opportunity to "rid themselves of uneconomic unwanted buildings".[73] These claims had merit. Heritage advocates' plans to preserve the civic centre, earmarked by Council for redevelopment prior to the earthquake, were dismissed by the Town Clerk as "bloody stupid".[74] The Council had its own plans for redevelopment of the city, articulated in the 1988 Newcastle CBD business plan, centred around a modern office, hotel and retail development and not heritage buildings. The earthquake offered an opportunity for its implementation. Similarly, the demolition of the damaged Royal Newcastle Hospital wings, which many considered unnecessary, helped fulfil the State Government's plans towards privatisation of hospitals and construction of the John Hunter hospital. But the advocates' argument

---

[71] "Conservationists concerned on demolition," *Newcastle Herald*, 2 January 1990, 6.

[72] Staff reporters, "Heritage fight as quake dust settles," *Sydney Morning Herald*, 1 January 1990, 2.

[73] Henry, "The battle for Newcastle," 107; Nathan Vass, "Row erupts as city hotel falls," *Newcastle Herald*, 6 January 1990, 1; Karen Davey, "Quake victims conned," *The Sun-Herald*, 7 January 1990, 1.

[74] Henry, "The battle for Newcastle," 108.

was also problematic as the city had in previous decades demolished several heritage buildings and neglected others (leaving them more vulnerable to earthquakes). To some extent Newcastle's self-identity as a heritage city was honoured more in rhetoric than in practice.

Council officer and Earthquake Coordinator, Harold Stuart, defended Council actions and declared the accusations of wanton destruction as the delusions of sensational reporting and residents with a political agenda. The decisions, he maintained, made economic sense.[75] But as in Christchurch, the rapid demolition of heritage buildings signalled a "return to the blank slate ideology of the mid-nineteenth century" in New Zealand,[76] a philosophy also characteristic of the colonial mentality in Australia. Newcastle's escalating controversy over preservation or rebuilding prompted NSW State Government intervention and the sending of a team of five experts, including engineers and New Zealand earthquake specialist, Bruce Shephard, to assess the damage.

On 2 January 1990 after receiving engineers' reports, Lord Mayor John McNaughton conceded that "there will be no wholesale destruction of the city", most buildings can be saved. McNaughton pledged, "no heritage buildings would be demolished".[77] However, the Council only recognised the heritage significance of buildings in the Hunter Regional Environmental Plan (Heritage), not the National Trust register. Among the 90 buildings approved for total demolition and 200 for partial demolition in February 1990, many that contributed to Newcastle's architectural character (e.g. suburban houses and Beaumont Street shops) were lost.[78] Those responsible later admitted that some demolitions had occurred prematurely, buildings could have been repaired.[79]

In contention were differing values and conflicted notions of place and identity. While Council wanted modern development, heritage advocates

---

[75] Stuart, "The Newcastle Earthquake," 18.

[76] Pickles, *Christchurch Ruptures*, 129.

[77] Bernard Lagan and Janet Fife-Yeomans, "NSW can't pay for rebuilding says Greiner," *Sydney Morning Herald*, 2 January 1990, 5.

[78] The Institution of Engineers, Australia, *Newcastle Earthquake Study* (Barton, ACT: The Institution of Engineers, Australia, 1990), 9.

[79] Henry, "The battle for Newcastle", 104–106. Pickles also notes unnecessary destruction of Christchurch heritage buildings with experts relegated to an advisory role. Pickles, *Christchurch Ruptures*, 25 & 137.

worked to retain the character of the past, keeping what was left in the face of so much lost. Behind both motivations lay the need to re-establish human control and a sense of normality as quickly as possible.

Brisbane too focussed on rebuilding and re-establishing control. In post-flood Brisbane the dominant narrative ran that the cruelty of nature would be overcome through human endeavour. Premier Bligh called on people's resilience to remake the city:

> as shattered lives are rebuilt, piece by piece, brick by brick, I want us to remember who we are, to take strength from that great Queensland spirit and conquer the challenge ahead of us.[80]

Emergency services, the army and state and local government rallied to clean, give aid to the desperate and rebuild. Flooded streets became an active hive of volunteers, labelled the Mud Army, helping the recovery. Lord Mayor, Campbell Newman urged residents: "Please get back to work". "We have to get this city cranking back into gear again. It's just as important to the overall recovery in Brisbane".[81] There was a more sinister undercurrent, humans should not be defeated by nature. Journalist Shane Rodgers clearly articulated this sentiment: if "we stand shoulder-to-shoulder", we can "assert our right to live and prosper".[82] Floods should not threaten capitalistic gain. While these physical responses restored the city's appearance, the psychological damage proved harder to address, with many facing property or employment loss, relationship strain or anxiety and more. A larger problem loomed at a societal level. Brisbane resident's sense of environmental security had been violated, the myth of living in a flood-proofed city destroyed.

Often ignored were the flood's anthropogenic causes, most notably the intense development of the floodplain. Poor land management policies by state and local governments for the benefit of property owners, government tax collectors, and powerful building and property industries

[80] Anna Bligh, "Message of support from the Premier," *Courier-Mail*, 19 January 2011, 36.

[81] Anna Caldwell and Sarah Vogler, "Chaos to confront workers as reality sinks in," *The Courier-Mail*, 17 January 2011, 2.

[82] Shane Rodgers, "Mud and Misery," in *The 2010–2011 Floods Crisis*, APN Special Commemorative Publication, February 2011, 5.

had substantially increased flood vulnerability.[83] But unwelcome discussions on human culpability threatened the status quo, the power relationships in a pro-development city and raised the unconscionable spectre of curtailing floodplain development. Instead focus turned to Somerset and Wivenhoe Dam, thought by many to be a panacea for floods. In reality, heavy rain and runoff had overfilled Somerset and Wivenhoe Dam's finite flood storage capacities, forcing water releases that inundated the city, but this was not what Brisbane residents wanted to hear.

An increasingly risk-adverse society searched for someone to blame. Within days of the flood, parts of the media and wider community framed the event as dam mismanagement, a positioning that successfully deflected attention from poor land-use practices and government fault. On 17 January the State Government established the Queensland Floods Commission of Inquiry, the examination of dam operations and operators included in its brief. Newspaper articles referring to the "great avoidable catastrophe", caused by dam mismanagement, shifted the Inquiry's focus to blaming the dam engineers.[84] As engineer Trevor Grigg noted, a strategy of blaming the bureaucrat re-enforced a dangerous view that "if only we get the operation of the dam right, we won't have any flooding of this river".[85] As state and local government rebuilt riverside boardwalks and ferry terminals and residents restored their damaged homes, they masked the flood hazard allowing a return to the imagined idyllic riverside city, with the dams continuing to provide the delusion of environmental security and control.

---

[83] See Margaret Cook, "Vacating the Floodplain: Urban Property, Engineering and Flood in Brisbane (1974–2011)," *Conservation and Society* 5, no. 3 (2017): 344–354.

[84] Hedley Thomas, "The Great Avoidable Flood," *The Australian*, 22–23 January 2011, 10; Margaret Cook, *A River with a City Problem: A History of Brisbane Floods* (Brisbane: University of Queensland Press, 2019).

[85] Quoted in Sophie Cousins, "Wivenhoe release could've prevented floods? Nonsense, say experts," *Crikey*, 16 February 2011, https://www.crikey.com.au/2011/02/16/wivenhoe-release-couldve-prevented-floods-nonsense-say-experts/.

Conclusion

In both Newcastle and Brisbane, residents were shaken, as disasters violated their sense of place and notions of environmental security. Novocastrians could no longer ignore the seismological risk of future earthquakes. But they could fight to preserve a sense of order and control. Authorities demolished damaged buildings to re-establish city activity. Others fought to protect and restore what the earthquake had not destroyed, culminating in a battle over heritage preservation. Heritage values are inextricably linked to notions of identity and character, building conservation intended to inform present and future residents about the past. The authorities' and heritage advocates' responses reflected an adherence to their own constructed perceptions of municipal character, feelings of belonging in a specific locale and a need to control their built environment. Debates on restoration or redevelopment avoided deep consideration of anthropogenic causes and obscured the human capacity to adapt.

In Brisbane, safe occupation of the riverbanks depended on dams withholding episodic floods characteristic of a sub-tropical climate. Many believed Wivenhoe Dam had the capacity to prevent floods; many still do. When flooding challenged this delusion, society clung to the hope that if the dams had been managed properly the flood would not have occurred. After the flood, the State Government altered the dam operations manual to adjust the dam release procedures, re-affirming the popular belief that dams would provide flood immunity.

In Newcastle and Brisbane the scramble to re-establish the status quo eroded the opportunity to question the anthropogenic causes of the disaster. Scholarship shows that where hazards are recognised as the product of political and social choices, the potential for hazard reduction and resilience is increased; the alternative leaves the city as vulnerable as before.[86] Destruction through natural disasters is regarded not in amazement but as a foreseeable risk against which precautions can be taken.

[86] Mark Pelling, "The Political Ecology of Flood Hazard in Urban Guyana," *Geoforum* 30 (1999): 249–261; Ben Wisner, Piers Blaikie, Terry Cannon, Ian Davis, *At Risk: Natural Hazards, People's Vulnerability and Disasters* (London: Routledge, 2003); Ben Wisner, "The Political Economy of Hazards: More Limits to Growth?" *Environmental Hazards* 2, No. 2 (2000): 59–61.

The reactions in Newcastle and Brisbane reflect an enduring notion that nature is external to the human environment, the human and non-human divide remains. Both city's self-identity is a construct, removed from its real environmental context of cities built in a volatile seismic area or on a floodplain. Neither earthquakes nor floods are preventable; they can only be managed through building codes, land-use plans and dams. Instead both cities rebuilt in situ, the delusion of environmental control survives.

What society needs is for people to be shaken *and* stirred to stimulate debate on the anthropogenic factors behind the disasters. In Newcastle, construction on alluvial soil and infill and the absence of earthquake building codes had contributed to the human dimension of the "natural" disaster. Dense construction on the floodplain and poor development codes had intensified Brisbane's flood hazard. Wider debate on the adaptability of human behaviour after the disasters may have spurred greater action and reduced both cities' vulnerability. By clinging to the past and myths of identity, environmental security and control, these communities are destined to suffer the same fate in the future.

Escaping Water: Living Against Floods in Townsville, North Queensland, from Settlement to 2019

Rohan Lloyd, Patrick White, and Claire Brennan

On 24 January 2019, Townsville's Mayor, Jenny Hill, triumphantly told local reporters that the city had reached an "important milestone" in its water history. Townsville was struggling through a severe drought and Hill proclaimed the beginning of a $215M pipeline designed to pump water to Townsville from distant sources. Yet, within one week of celebrating the installation of the drought-proofing pipeline, Hill was managing an evolving disaster. Persistent torrential rain had caused flash flooding in low-lying areas, and by 3 February, Townsville's dam water had swollen from a mere 15 per cent capacity to over 250 per cent. The dam's gates were thrust open, releasing around 1900 cubic metres of water per second into the swollen Ross River below. What was initially a drought-breaking weather event quickly became a catastrophic flood. The contrast between drought and deluge, often apparent in Townsville, and the events of 2019 reflected a 150-year struggle between the natural environment and the city of Townsville.

R. Lloyd (✉) • P. White • C. Brennan
James Cook University, Townsville, QLD, Australia

S. McKinnon, M. Cook (eds.), *Disasters in Australia and New Zealand*,
https://doi.org/10.1007/978-981-15-4382-1_6

During the flooding of 2019, many homes were left unliveable and much of Australia's most populous northern city became an urban archipelago.[1] The disaster was the largest and most destructive flooding to have occurred in the city since colonial settlement, and it prompted the Insurance Council of Australia to declare Townsville's region the most flood-prone in Australia.[2] The intense rainfall (approximately two metres in two weeks) also relieved the effects of a long-standing drought and filled the city's dam. Initially, the community welcomed the replenishing effects of the intense rainfall event: for several years before, Townsville's local government had enforced water restrictions and encouraged limiting residential and commercial use.[3] Fresh water was in short supply and a string of drier-than-normal "wet seasons" brought the issue of water to the forefront of local politics. In response to drought, the Queensland government and the Townsville City Council (Council) committed in 2017 to "guarantee the city's future water supply" through a $215M construction project.[4] The floods, however, provided residents with a stark reminder of their climatic and environmental realities.

Established in 1864–65 on the Ross River plain, Townsville is located within Queensland's drought-prone, dry tropics (Fig. 1). It experiences an average annual rainfall of 1143 mm. The bulk of this rain falls within the northern "wet season" between November and April and is associated with monsoons, tropical storms and cyclones. Despite the damage which can often accompany them, the city relies upon extreme rain events for its urban water supply. Townsville is also reliant upon, and at the mercy of, the Ross River catchment system. The city sprawls across the Ross

[1] "Northern Australia" is defined here as the continental area north of the Tropic of Capricorn.

[2] Felicity Caldwell, "Townsville is Australia's most flood-prone area," *Brisbane Times*, 20 March 2019: https://www.brisbanetimes.com.au/national/queensland/townsville-is-australia-s-most-flood-prone-area-20190320-p515rj.html.

[3] See Bettina Warburton, "Council Confirms water restrictions are here to stay," *Townsville Bulletin*, 2 May 2018: https://www.townsvillebulletin.com.au/news/council-confirms-water-restrictions-are-here-to-stay/news-story/a9bf046396d6350dcceb2844d7ba51d2 and Tony Raggatt, "Council says restrictions are needed to maintain water supplies," *Townsville Bulletin*, 13 December 2018: https://www.townsvillebulletin.com.au/business/council-says-restrictions-are-needed-to-maintain-water-supplies/news-story/b67cfd9d85fee3c28be70601b676d3cf.

[4] Townsville City Council, *Breaking Ground on $215 million water pipeline*, 20 November 2017: https://www.townsville.qld.gov.au/about-council/news-and-publications/media-releases/2017/november/breaking-ground-on-$215-million-water-pipeline.

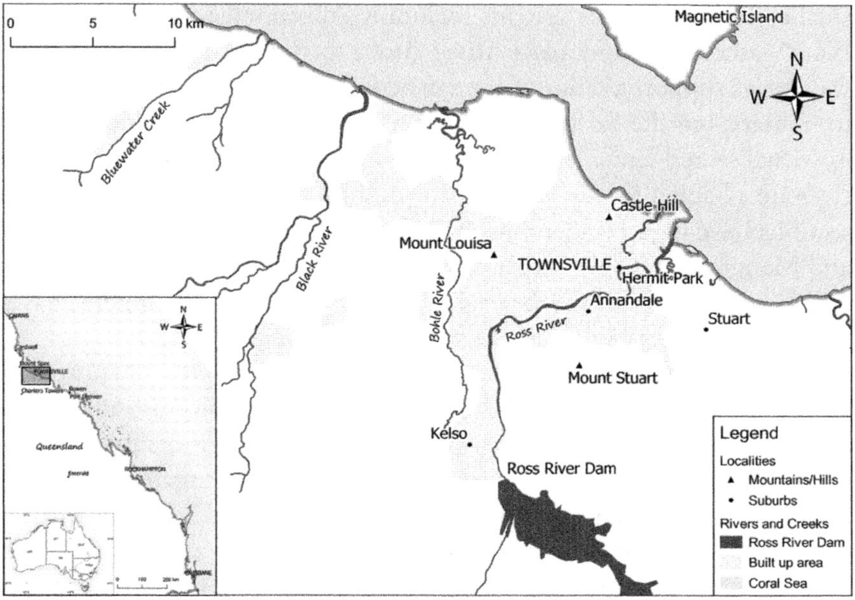

Fig. 1 Map of key locations in Townsville and surrounds

River catchment, and its suburbs stretch into other watersheds. These watersheds have supplied fresh water to Townsville residents since colonial settlement and the Ross River itself has been altered by the city. It has been dammed 26.4 km upstream from the mouth (the Ross River Dam is in the outlying suburb of Kelso) and the construction of a series of weirs has changed it from an ephemeral river into a permanent presence, albeit often as a series of ponds rather than a flowing waterway.[5] The Ross has a number of small distributaries, but it is the Ross River itself which flows through the city's suburbs before reaching the ocean.

In recent years the sprawl of Townsville's residential area to the north has made it a two-river city, and to the north of the Ross catchment lies Townsville's other river system—the Bohle. The Ross and the Bohle share hydrological connections through surface, ground and storm water. Furthermore, the city has a number of satellite suburbs that incorporate

[5] Peter Bell, *History of Ross River* (Thuringowa: Thuringowa City Council, 2003).

smaller creek and river systems including Alligator Creek to the south, Blue Water Creek and Black River (both to the north-west). All these waterways support a range of human activities, from pastoral, to mining, to military, but the Ross in particular is an important recreation zone for Townsville's residents. In addition, the upper freshwater sections of the river are a habitat for a variety of fish and birds, while the lower estuarine sections have high populations of the popular sporting fish Barramundi and Mangrove Jack. Yet, despite residents' familiarity with water and the visible controls that have been placed on its main river, the city has a problematic history of balancing consumption and flood.

Water, drought and flood have long attracted the interest of Australian environmental historians, and Townsville's position as an Australian tropical city gives its flood history a significant, though yet incomplete, place in this historiography. While within Queensland Margaret Cook has undertaken a detailed examination of urban flooding, her work focuses on attempts to control the Brisbane River and makes a contribution to the history of Brisbane city.[6] A similarly tightly focused study by Scott McKinnon has analysed the role of flood as both a part of Brisbane urban identity and a peculiar lacuna in its collective memory.[7] In contrast, the histories of water written about regions further north analyse development schemes that seek to use "wasted" northern waters for irrigation, rather than examining urban waterways. Both Russell McGregor and Lesley Head have used the Ord River Irrigation Schemes to examine the ways in which political imperatives for northern development have overridden economic concerns. In the post-war period, continuing a long history of fascination with northern irrigation schemes, concerns about an "empty north" made the creation of irrigation infrastructure imperative in government attempts to extend closer settlement in Australia's tropics.[8] In contrast, the management of water within existing northern

---

[6] Margaret Cook, "Damming the 'Flood Evil' on the Brisbane River," *History Australia* 13, no. 4 (2016): 540–556; Margaret Cook. "'It Will Never Happen Again': The Myth of Flood Immunity in Brisbane," *Journal of Australian Studies* 42, no. 3 (2018): 328–342.

[7] Scott McKinnon, "Remembering and forgetting 1974: the 2011 Brisbane floods and memories of an earlier disaster," *Geographical Research* 57, no. 2 (2019): 204–214.

[8] Lesley Head, "The Northern Myth Revisited? Aborigines, environment and agriculture in the Ord River Irrigation Scheme, Stages One and Two," *Australian Geographer* 30, no. 2 (1999): 141–158;

cities has not yet been explored by historians, and Townsville's 2019 floods offer an opportunity to scrutinise flood history on Australia's suburban frontier.[9]

This chapter explores Townsville's water history, its experiences of drought and flooding, and its responses to these extremes since the 1860s. We argue that the city's ambivalent position as a suburban frontier plays out in media coverage and in local responses to its challenges with water. In the media, images of crocodiles in urban floodwaters, and interviews with straight-talking locals, portray Townsville as a frontier settlement. Yet, for residents of the city, water management schemes and disaster responses are expected to sustain a standard of living matching that found in the suburbs of Australia's major cities. Acceptance of monsoonal deluges and the big-wet sits awkwardly with a lack of acknowledgement of the long dry season and the destruction brought by extreme weather events. Unpredictable cycles of below-average rain followed by extreme wet weather have informed locally unique ways of processing, memorialising and rebuilding from these disasters. While stoic responses to major weather events have become part of the city's collective identity, the community's acceptance of recurrent but unpredictable major weather events has not led to the development of a sustainable flood memory.[10] Understanding of flooding in the city is poor, and the possible depths and

---

Russell McGregor, *Environment, Race, and Nationhood in Australia: Revisiting the Empty North* (New York: Palgrave Macmillan, 2016), 172–178.

[9] The term "suburban frontier" is taken from Lyndon Megarrity, who coined the term while researching his book *Northern Dreams: the politics of northern development in Australia*, Pers Comm. 16/6/2019.

[10] The concept of "flood memory" has been used with increasing frequency in studies of memory and flooding. The concept has been developed further in Lindsey McEwen, Franz Krause, Owain Jones, et al. "Sustainable flood memories, informal knowledges and the development of community resilience to future flood risk," in *Flood Recovery, Innovation and Response III*, ed. D. Proverbs, S. Mabretti, C.A. Brebbia, et al. (Ashurst: WIT Press, 2012), 253–264; Joanne Garde-Hansen, Lindsey McEwen, Andrew Holmes, Owain Jones, "Sustainable flood memory: Remembering as resilience," *Memory Studies* 10, no. 4 (2017): 384–405; and Lindsey McEwen, Joanne Garde-Hansen, Andrew Holmes, Owain Jones and Franz Krause, "Sustainable flood memories, lay knowledges and the development of community resilience to future flood risk," *Transactions of the Institute of British Geographers* 42, no. 1 (2017): 14–28. Similarly, Donald A. Wilhite's description of a "hydro illogical cycle" can be seen as a critique of failures of "drought-memory", "Breaking the Hydro-Illogical Cycle: Changing the Paradigm for Drought Management," *EARTH Magazine* 577, no. 7 (2012): 71–72.

intensities are not well recognised. The residents of Townsville, and their reactions to flooding, offer lessons in both accepting and ignoring the local environment while constructing an urban settlement within the tropics.

Settling on a Floodplain

Townsville's early history was partially defined by water and attempts to modify the natural environment. European settlers have battled Townsville's natural environment and attempted to deny its tendency to flood since the first residents and labourers began displacing the local Wulgurukaba and Bindal peoples. Access to water was a decisive feature in the siting of Townsville by settlers, but it was the ocean (not fresh water) which determined the city's location on the banks of the Ross Creek in Cleveland Bay. The settlement became an important port, which served pastoral, mining and (later) sugar interests. While the area's shallow waters and the sediment left by floods made the port navigationally inferior to the alternatives available at Bowen to the south and Cardwell to the north, Townsville's superior access to the north Queensland hinterland elevated the port's commercial appeal and later incentivised flood mitigation.

Early settlement in Townsville hugged the banks of Ross Creek, a distributary of the Ross River, which wound its way towards Cleveland Bay across a low-lying plain bordered by Mount Stuart and Castle Hill. The site was selected by Andrew Ball, in a survey undertaken on behalf of John Melton Black. The settlement's name reflected the interests of Robert Towns, whose pastoral properties Black managed. Black held concerns about the distance and obstacles that lay between Towns' properties and Port Denison, the nearest established port approximately 200 km away at Bowen, and instructed Ball to conduct the survey.[11] When Ball encountered the Ross in April 1864, it was in flood. Undiscouraged, Ball

[11] See Glen Lewis, *A history of the ports of Queensland: a study in economic nationalism* (St Lucia: University of Queensland Press, 1973); H.J. Taylor, *The History of Townsville Harbour—1864–1979* (Brisbane: Boolarong Publications, 1980).

reported favourably on the area's suitability for a port and encouraged by the prospect of a new port and motivated by the potential for profit from residential and commercial land sales, Black and Towns committed to develop a site on the banks of Ross Creek. A town was quickly established: by November 1864, clearing of mangroves had commenced and the sites for wharves and buildings had been picked out.[12] By February 1866, the settlement's population had risen to over 400 people and the Municipality of Townsville was proclaimed.[13]

The new settlement soon experienced the vagaries of Townsville's rainfall, and dry periods were broken by unpredictable flooding. Large floods in 1870, 1881, 1884, 1890 and 1892 demonstrated how heavy rain and tidal surges, particularly during cyclones, caused destruction to the developing city and its vulnerable port.[14] People and governments rebuilt Townsville, however, largely because of the wealth of its hinterland and the security afforded by lengthy dry periods. And urban expansion continued among the river's flood plain.[15] In her study of settlers on the Hawkesbury-Nepean River in early New South Wales, historian Grace Karskens argues that settlers there accepted flood as irregular, destructive events that did not undermine the value of productive soil. She claims such views persisted until at least the late 1990s.[16] Such acceptance of flooding as part of the natural order is unusual in Australian history, and as early as the 1890s, civic and economic leaders in Townsville began to investigate the flood mitigation potential of hydro-engineering. Early mitigation proposals focused on diverting the upper Ross River into the Bohle River. However, ideas of redirecting the flow of Townsville's major

---

[12]Townsville 100, Townsville City Council, 1964—see North QLD special collections.

[13]"Municipality of Townsville Proclaimed," *Queensland Government Gazette* vii, no. 22 (17 February 1866): 211.

[14]"Townsville," *Warwick Examiner and Times*, 2 April 1870, 2; "Townsville," *Queensland Times, Ipswich Herald and General Advertiser*, 29 March 1881, 2; "Townsville," *Brisbane Courier*, 2 April 1881, 6; "Heavy Floods in the North," *Brisbane Courier*, 25 February 1884, 5; Hector Holthouse, *Cyclone* (Adelaide: Rigby, 1971), 41–45.

[15]"Aboriginalities," *Bulletin*, 3 August 1901, 14. While the article exhibited horror that special residential leases were granted to a number of non-white settlers, particularly Chinese, at sites along the Ross River in 1901, the report evidences continued urban expansion along the river's flood plain despite the city's historic flooding.

[16]Grace Karskens, "Floods and Flood-mindedness in Early Colonial Australia," *Environmental History* 21 (2016): 324.

river away from its natural delta, adjacent settled areas and burgeoning port were discredited by Charles Napier Bell, a Scottish-born hydro-engineer with experience derived from projects in Brazil, Argentina, Russia and New Zealand. Bell worked on hydro-engineering projects throughout the Australasian colonies, and in 1898, he advised Townsville authorities that diverting the Ross would provide only minimal relief from flooding.[17]

In her study of Brisbane's dams, Cook argues that frequent and long-lasting drought was a key factor retarding development of flood-mitigating dams in Queensland's capital. In the absence of significant flooding events, the need to secure a water supply for the city blunted the urgency of flood mitigation.[18] Water scarcity also influenced development in Townsville. Urban water first came from a series of wells, which dotted the townscape, and in the early twentieth century, a series of weirs (Black, Gleeson and Aplin) was established along the Ross River to improve water storage. Even with these measures, Townsville quickly became parched when the rains failed. The *Courier-Mail* declared that Townsville's 1935 "wet season" was the driest in the past 66 years and water restrictions were common. But the relationship between drought, flood and confidence in the future of the city was complex. Lengthy droughts led to complacency about the dangers of Townsville's position and encouraged development in low-lying areas. Flood, when it came, extinguished the doubts about future prosperity that drought had allowed to smoulder and in turn encouraged urban and industrial expansion.

[17] See "Obituary. Charles Napier Bell, 1835–1906," *Minutes of the Proceedings of the Institute of Civil Engineers* 164 (1906): 401–403; Queensland Irrigation and Water Supply Commission, "Preliminary report on investigation of flood mitigation and water resources development" (Brisbane: Queensland Government, 1965). Copy of report held by the James Cook University's Eddie Koiki Mabo Library Special Collections.

[18] Cook, "Damming the 'Flood Evil'," 541.

Post-war Floods and Ross River Dam

Events in 1946 in Townsville raised city residents' awareness of flood, although the local discourse around water only slowly shifted from one which focused on supply to one that included flood mitigation. The year 1946 was one of Townsville's wettest summers on record, and it included two major floods. The first followed heavy rain in January and February, and it caused Townsville's largest flood in 40 years. A levee on the south side of Aplin's Weir burst spilling 400,000,000 gallons of water into what was then unpopulated but is now the large suburb of Annandale.[19] Despite the extensive flood, it was damage to the weir and the lost opportunity to capture valuable water which dominated the discussions in Townsville immediately after the event. Typically, the *Townsville Daily Bulletin* was concerned that the breach was "shortening … the supply by 100 days" as daily "4,000,000 gallons" rushed past the wall and out to sea.[20] The city's experiences of lengthy drought meant that escaping water, even if alleviating localised flooding, could scarcely be tolerated.

But only weeks later, on 2 and 3 March, the city experienced another significant weather event. On that weekend, huge downpours fell across the Ross River catchment, spreading "a sea" across the entire city.[21] Streets became torrents, two houses in the suburb of Hermit Park were swept away, four people lost their lives, and there was widespread damage to businesses, homes and infrastructure. The *Townsville Daily Bulletin* reflected on the significance of the flood within the history of the city and explicitly discussed the development of areas within the flood's reach:

> There is no record here of previous rain like that experienced on the morning of March 3 and the records show that such torrential downpours are rare throughout the whole of Australia … Townsville has never suffered such devastation from a flood, though old residents can recall at least one similar visitation. At that time, however, we were a very small community, and there was little settlement in the flooded areas. This time it was differ-

19 "Townsville Cyclone," *Australian Worker*, 13 February 1946, 12.

20 "Breach in Aplin's Levee May Drain Weir of Water," *Townsville Daily Bulletin*, 12 February 1946, 1.

21 "Highest Ross Flood in 50 Years?", *Townsville Daily Bulletin*, 4 March 1946, 1.

> ent. Large residential areas lay in the path of the flood waters, and unfortunately the flood was attended by loss of life, while many homes were damaged and many of our citizens striving to secure homes for themselves, face ruin.[22]

The editorial asserted there was "much to learn … in minimizing loss in the future, even though there may not be such another [flood] for many years to come".[23] While there was widespread acceptance that there had been unprecedented rainfall, there was also public criticism of the role the water-saving weirs played in exacerbating flooding. The city's engineer, Felix Howard Brazier, defended the weirs. He argued that the construction of weirs had been necessary to ensure water supply for the city (he had himself overseen the construction of Aplin's Weir in the late 1920s) and that their value in this regard would only increase with increasing population.[24] As for flood mitigation, Brazier argued that Aplin's weir had played a useful role in reducing flood damage as the levee on the northern bank (which had been repaired after the February flood) had forced flood waters over the southern bank into an undeveloped area.[25]

The 1946 floods proved significant in the development of a Townsville flood memory. The second flooding event forced the residents of Townsville to recognise their limited experience of flood damage. And that realisation was recorded as Townsville residents for the first time captured material memories of flood in the form of photographs. Images of the flooded city were published in the *Townsville Daily Bulletin*, and reports of the flooding mixed anecdotes of residents' experiences with swirling prose about the flood-waters' movement through the city.[26] The 1946 floods, particularly the second event, became a high-water mark for the city. In flooding events in 1950, 1951 and 1953, the two 1946 floods

[22] "The Ross Flood," *Townsville Daily Bulletin*, 14 March 1946, 2.

[23] "The Ross Flood," Townsville Daily Bulletin, 14 March 1946, 2.

[24] Geoffrey Cossins, ed., *Eminent Queensland Engineers Volume II* (Brisbane: Institution of Engineers, 1999), 23; Townsville City Libraries, "Heritage Services Information Sheet Number 6: Ross River Weirs": https://www.townsville.qld.gov.au/__data/assets/pdf_file/0005/21011/HERITAGE-INFO-SHEET-6-Ross-River-Weirs.pdf [accessed 26 March 2019].

[25] "Ross River Weirs," *Townsville Daily Bulletin*, 22 March 1946, 2.

[26] See "Highest Ross Flood in 50 Years?" and "South Townsville Flood Scenes," *Townsville Daily Bulletin*, 7 March 1946, 3.

became important framing devices for remembering.[27] These two were modern floods in a modernising city; they were able to be captured and memorialised and form part of an urban collective memory. The growth of the city had undoubtedly contributed to an increase in flood-related damage, but this alone does not explain the lingering awareness of the 1946 events: there had also been an increase in the infrastructure of remembering.

Despite the material awareness and subsequent reference to 1946, it did not mark a permanent shift in the city's conception of floods as part of the climate, rather than as extreme and unprecedented events. McKinnon argues that while Brisbane's 1974 flood became a reference point for long-term residents, it also became "securely located in the past".[28] In Townsville in 1946, as in Brisbane 30 years later, residents moved from dealing with the damage caused by the flood to discussing securing a reliable water supply, without pausing to consider how to deal with future floods. Drought, rather than flood, directed public discussion, and subsequent, smaller flooding events were celebrated for their drought-breaking status despite the experiences of 1946. After the 1953 floods, despite the anxieties it caused, the *Townsville Daily Bulletin* editorialised:

> North Queenslanders in general should be in great heart with the bounteous rains which have fallen. From a country which, in most areas, has been parched almost beyond description, it has been turned into one which is again full of promise, replete with that moisture so necessary for all living things, be they human, animal or vegetable.[29]

The slow anxiety of dealing with yearly urban-droughts often relegated the intense, and immediate, damage associated with floods.

[27] See "Ross River Overflows, Floods Low Lying Areas," *Townsville Daily Bulletin*, 9 March 1950, 1; "Floods, Gales Wreaks Havoc in the North," *Townsville Daily Bulletin*, 12 January 1951, 1; "Floods in Townsville," *Townsville Daily Bulletin*, 16 January 1953, 1.

[28] McKinnon, "Remembering and Forgetting," 205.

[29] "North Will Benefit in Large Measure," *Townsville Daily Bulletin*, 15 January 1953, 2.

Flood Mitigation and Water Supply

Townsville Council's approach to flood mitigation was consistent with other local councils in Australia at the time.[30] Rather than address flooding issues through changes in human interactions with the flood plain, the Council set about investigating options for controlling and engineering the Ross and its flood flow to protect the existing path of development. However, the Council's outlook was grounded by an acceptance of uncertainty around future weather events, the limits of memory and experience, and dependency on external expertise to improve and shape water and flooding infrastructure.

In 1951, the Council commissioned the Brisbane engineer, William James Reinhold, to investigate the flooding of the Ross River. The Council requested Reinhold to assess the merits of two specific schemes: the diversion of the Ross' flood flow into the Bohle River (revisiting the earlier proposal) and the building of a levee bank to shield the city further down the river. The proposed levee was to be 3 m high, 3 m wide at the top and nearly 20 m wide at its base. It would stretch from the banks of the Ross to Mount Stuart on one side (a distance of 1 km) and to Mount Louisa on the other (4 km).

Reinhold's report dismissed the viability of either option. Like Bell in 1898, Reinhold argued that "a significant reduction of major flooding is impracticable". Instead, he proposed a diversion channel that would run from the Ross to the sea which would allow the reclamation of land on the northern bank and minimise moderate flooding. Overall, Reinhold favoured the prosaic technologies of dredging and channel improvement to avoid the necessity of rerouting roads and rail networks which would follow the levee or rerouting options.[31] William HR Nimmo, Queensland's Commissioner of Irrigation and Water Supply, agreed with Reinhold's conclusions and suggested that diverting the Ross' flood flow to the Bohle, "except by means of works which will provide complete control", would be dangerous, excessively expensive and likely to cause a permanent diversion of the Ross River to the Bohle Valley. Despite these

[30] Cook, "Damming the 'Flood Evil'," 540–541.

[31] "Ross River Flooding," *Townsville Daily Bulletin*, 20 May 1952, 3.

criticisms, during the 1950s, the Council periodically revisited plans to divert the Ross River. In 1953, the Council task the Civil Engineering Department at the University of Queensland (UQ) with creating a "scale model of Ross River topography to prepare a flood prevention plan".[32] In 1957, the Department reported that the Ross could be diverted to the Bohle above Black Weir, but flood-modelling did not include testing of the effect of subsequent flooding on the city located further down the river. The option of cutting a channel from the southern bank of Aplin's Weir towards the ocean was also considered, but this, they claimed, would likely result in the flooding of the "rapidly growing industrial area of Stuart".

While it was discussing plans for flood mitigation, the Council was also struggling to manage the city's water supply. Increased development of North Queensland's mineral and agricultural industries contributed to population growths for its major port cities; between 1947 and 1961, Townsville's population grew by 50 per cent, intensifying the pressure to secure a reliable and adequate water supply. In 1954, the Council completed the construction of a pipeline to Crystal Creek at Mt Spec; 5 years later the Council supplemented this supply by constructing a dam on the Paluma Range. The cost of these projects, nearing £900,000, was very close to the anticipated cost of diverting the Ross and public debate linked the two concerns.[33] The *Townsville Daily Bulletin* was cautious in its evaluation of the plans to divert the Ross, reminding readers:

> It is well to remember that Townsville has to look to the Ross from two angles. In the first place, it has been the source for the past 50 years of the city's water supply and, secondly, it has to be viewed from the hazard perspective of what it can do in times of floods. The Ross appears to be the main source of Townsville's water supply for many years to come ... Therefore, the water needs value of the Ross must be considered prior to any drastic alterations to the course.[34]

[32] "Townsville Flood Crisis Not Yet Passed," *Cairns Post*, 16 January 1953, 1.

[33] This is roughly A$31 million in today's money.

[34] "Diversion of Ross Waters May Take Years to Put into Effect," *Townsville Daily Bulletin*, 20 November 1954, 2.

Despite such public appreciation of the ongoing role of the Ross in supplying Townsville with water, in April 1959, the Council requested UQ to undertake additional tests with a view to determining the size and nature of diversion structures required to ensure "that no flood would flow down the Ross River".

Concerns about the Ross as a source of flood were again made clear when the Coordinator General of Public Works requested that the Irrigation and Water Supply Commission take over the examination of the proposals to divert the Ross. In 1959, the Commission recommended against the diversion proposal. The Commission suggested that the Council consider the construction of a dual-purpose dam (a solution already employed in Brisbane with the construction of Somerset Dam) on the upper section of the Ross to provide flood mitigation and augment water supply. In 1965, in response to Townsville's increasing population and growing urban sprawl heightening fears around serious flooding hazards, the Commission made a formal recommendation to the Council to abandon the diversion plan and take immediate action to construct the Ross Dam. Construction of the dam commenced in 1971 and was completed in 1974.

Modern Townsville

After the construction of the Ross River Dam, Townsville experienced few major riverine flooding incidents. Those flooding events made apparent the vulnerability of the city's older suburbs, located close to the mouths of the Ross Creek and Ross River. In those areas, the flat topography means there are limited stormwater surcharge paths which are compounded by undersized drainage paths.[35] Before construction of the dam, Townsville's flooding concerns centred on riverine flooding; post-dam, they were limited to suburbs vulnerable to storm surge and localised catchment flooding. The vulnerability of Townsville to such flooding was made apparent on 10 January 1998. That night Townsville recorded

[35] Townsville City Council, *Townsville Flood Hazard Assessment Study, Phase 3 Report Vulnerability Assessment and Risk Analysis* (Townsville City Council, Townsville, 2005), 24.

549 mm of rain during the city's wettest 24-hour period. That rain was associated with ex-tropical cyclone *Sid*, which had already brought heavy rains and caused flooding north of Townsville. The damage from the flood was significant: one man lost his life; up to 100 residences were inundated; and Townsville's satellite communities of Black River and Bluewater sustained extensive damage including the washing away of eight homes. Townsville's foreshore, known as the Strand, was badly eroded, and on Magnetic Island (a short ferry ride from Townsville), a landslide caused major damage to a resort complex.

As with the 1946 event, the local media quickly started memorialising the 1998 flood. The *Townsville Bulletin* provided up to 5 pages of flood-focused news stories a day for the next fortnight, including a 4-page lift-out on the 1946 floods, and a final 16-page lift-out which, due to the initial run of papers selling out, was republished two days later.[36] Additionally, North Queensland Ten News produced a one-hour documentary called *The Flood: January '98* which aired locally and was then sold on videotape to raise money for the Townsville Flood Appeal. While the Bureau of Meteorology had warned the city that bad weather was predicted for the weekend, the extent of the event had not been forecast. A meteorologist in *The Flood* explained that extreme events "are very hard to forecast because of their nature, because they're so unusual, and because they develop … it develops right over you. It just happens". Despite the flood's association with ex-Cyclone *Sid*, a local radio personality named the event the Night of Noah, on the grounds that no one expected the flood, just like "no one expected Noah".[37]

While the coverage of the disaster naturally emphasised the event's size and the community's capacity to respond, it also hinted that things could have been considerably worse. The heaviest rainfalls had fallen just north of the city, and while the townships of Bluewater and Black River received substantial damage, those areas closer to the Bohle River, reports

[36] The original lift-out was published on 27 January 1998 and the republishing occurred on 29 January 1998.

[37] "Night of Noah: 20 years since record flood devastated Townsville," *Townsville Bulletin*, 10 January 2018. Alliteration made the name attractive despite the lack of accuracy in its Biblical reference.

suggested, "may have been very lucky".[38] A *Townsville Bulletin* editorial even instructed residents to count their "lucky stars" and proclaimed that

> perhaps now all those people who wonder wistfully what a cyclone would be like will have had their curiosity satisfied. What hit Townsville on the weekend was "only" a tropical low, not strong enough to rank in the lowest cyclone category. A few weeks ago Townsville was rejoicing in the wet weather, reveling in an infrequent soaking. The novelty has gone now that the clouds have rained tragedy upon us. It's time for sunshine.[39]

Media coverage quickly turned to those residents who were taking advantage of flood-aid, criticisms of the flood-warning systems and frustrations with insurance claims.

Considerably after "Noah", triggered by two close-calls with significant Tropical Cyclones (Larry in 2006, Yasi 2010), the city's disaster management strategy turned to tackling storm surges and cyclones. The city disseminated maps and educational materials about the risks of storm surges, and a Category-5 suitable cyclone shelter was built within the suburb of Heatley. While heavy rains would fall causing localised flooding, it was considered "unlikely" that Townsville was vulnerable to a major flood. Indeed, a 2009 Risk Management study of Townsville stated that while floods can be "potentially damaging" events, "the more frequently occurring levels of flood produce inconvenience rather than disaster. The inconveniences and losses brought by flooding are, to a large extent, offset in many rural areas by their replenishment of soil moisture and surface water".[40] Big-wets were welcomed reprieves from long dry-spells.

The material memory of the Night of Noah was almost exclusively provided by the local newspaper. In 2018, the *Townsville Bulletin* published a 20-year anniversary recap of the "Night of Noah". Over the 20

---

[38] D. King and B. Girling-King, "Townsville Thuringowa Floods: January 1998 Business and Infrastructure Post Disaster Survey," Report Prepared for Department of Emergency Services (Centre for Disaster Studies, James Cook University, 1998), 2.

[39] "Tragedy from the clouds," *Townsville Daily Bulletin*, 12 January 1998, 8.

[40] Townsville City Council, *Townsville City Natural Disaster Risk Management Study Stage 1: Draft Final Report* (Adelaide: The Institute for International Development, 2009), 141.

years since the flooding, the framing of the event had shifted. In the first instance, it was a disaster of biblical scale. It was damaging and destructive. This was "bad" water; it entered homes and disrupted the material security that they represent. Yet, time eroded some of this negativity. In later accounts, the Night of Noah almost took on the mantle of "good water".[41] While the reprinted images were diverse, the written elements of the reports emphasised the response of the community, and the antics and anecdotes of locals trapped in hotels, restaurants, shopping car parks and birthing suites. While the flood remained a hazard, its role in negatively impacting the private lives of others was forgotten in the remembering. In 2018, and in the weeks before the 2019 flood, the city's water problems were not with flood, but drought.

And so the Townsville flood of 2019, for all its seeming novelty to those caught in it, falls within the pattern of previous flood events. It was an extended and damaging flood that resulted in the deaths of four people and the inundation of several thousand homes and businesses, but the response to it revisited the city's obsessions with drought, resilience and its refusal to accept flood as an unavoidable fact of life. Tom Griffith's analysis of the 2009 fires in the Mountain Ash forests of Victoria and Cook's analysis of the response to the 2011 Brisbane flood, both note the inability of the affected communities to accept cyclic destruction as an unavoidable consequence of living in certain environments.[42] Instead, both disasters were popularly explained as resulting from exceptional circumstances and avoidable mismanagement. Greg Bankoff's analysis of disaster as a western construct, in which he distinguishes unavoidable natural hazard from the human experience of disaster, perhaps holds the key to understanding this difficulty. Bankoff argues that residents of the first world tend to see disaster as something associated with the poverty and ill-preparedness of the third.[43] Reversing this vision hints that those

[41] Gordon Walker, Rebecca Whittle, Will Medd, Marion Walker, "Assembling the Flood: Producing Spaces of Bad Water in the City of Hull," *Environment and Planning A: Economy and Space* 43, no. 10 (2011): 2304–2320.

[42] Cook, "'It Will Never Happen Again'"; Tom Griffiths, "An Unnatural Disaster?: Remembering and Forgetting Bushfires," *History Australia* 6, no. 2 (2009): 35.1–35.7.

[43] Greg Bankoff, *Cultures of Disaster: Society and Natural Hazard in the Philippines* (London: Routledge, 2002), 5–6.

who consider themselves residents of the first world may not be culturally equipped to accept disaster as part of their quotidian experience.

This uneasy surprise at nature sometimes still being ascendant in first world cities can be perceived in the reporting of the Townsville flood. In particular, images of crocodiles being present in Townsville's floodwaters were widely reproduced and gained international exposure. The reporting of sightings of crocodiles pushed out of their normal abodes varied, but much of it demonstrated surprise at tropical nature being at home in Australian suburbia. Townsville's urban crocodiles are of the freshwater variety not considered hazardous to humans, a point made repeatedly by the local Council through its online disaster dashboard. Ironically, the crocodile images picked up by local, national and international media were mostly drawn from a Facebook page normally used by residents of the leafy suburb of Annandale to alert each other to incidents of car theft, property crime, lost pets and other suburban concerns. The flood restricted the movements of the page's contributors as well as the commission of petty crime in the suburb, and members of the "Annandale Staying Aware" Facebook page instead posted images from the Ross and the parklands surrounding it as it flows through their suburb. Despite their seemingly apocalyptic presence in flood reportage, freshwater crocodiles are often seen along that stretch of the Ross River. Their notoriety during the 2019 flood points to concerns with Townsville's position as a suburban frontier and a desire to represent catastrophic flood as an unprecedented event; crocodiles within Townsville's residential areas were not the inversion of natural order they were portrayed as, and neither was the flood that brought them media attention.

Townsville's place in the tropics might add crocodiles to the usual experiences of irregular drought and flood caused by Australia's highly variable rainfall, but in many ways, the city's story is one familiar from other metropolitan areas, particularly Brisbane. What sets the experiences of Townsville apart is the location of its suburbs in tropical Australia. Here the myth of engineering overcoming natural hazard has particular appeal to a population looking to experience suburbia on Australia's northern frontier. Here, the common Australian preoccupation with managing water resources to ensure metropolitan comfort intersects with northern histories of grandiose irrigation schemes meant to promote agricultural settlement in frontier regions. In the suburbs of Townsville, the manifestations of flood are not as uncommon as suburban residents choose to believe.

The 2019 flood was represented as a 1 in 1000-year event, but climate change makes it likely another significant flooding will occur. However, drought is also an aspect of climate change, and a familiar foe. The Council has maintained flood support for those impacted, but its commitment to drought management stretches further into the future. Since the flood, apart from those who are still living with its impact, for many it has faded into memory; the water has vanished, and the obvious impacts of its destruction have too. What was a major public disaster has now shrunk and become concealed by private efforts towards recovery. Meanwhile, the Ross River Dam levels have reduced during another long dry season, and the pipeline construction continues. Together with a recent city-wide "Get Water Smart" programme, these initiatives show that water security is a more pressing local concern than deluge.

Acknowledgements Patrick White received support for this research from an Australian Government Postgraduate Award and a James Cook University's College of Arts Society and Education's MRF Competitive Funding Grant.

A Perfect Storm

Ian Townsend

Cape Melville, about 450 kilometres south of Thursday Island on the east coast of Cape York Peninsula, juts so far out into the Coral Sea you can watch the sun rise and set over the water. The far western shore of Princess Charlotte Bay is 80 kilometres away, well over the horizon. Closer, about 30 kilometres east, is the edge of the Australian continental shelf. Here the sea floor rises abruptly, squeezing the Great Barrier Reef into a narrow strip of relatively shallow water (Fig. 1). In the late 1800s, the area was the site of a thriving commercial pearl shell industry, and in March 1899, as many as 1000 people aboard mostly Thursday Island-based fleets were working the shell beds near Cape Melville.[1]

[1] "Hurricane in the North," *Brisbane Courier*, 10 March 1899, p. 5. Pearling fleet owner James Clark said, "The crews employed on these [company] vessels would number in all not far short of 1,000."

I. Townsend (✉)
School of Historical and Philosophical Inquiry, University of Queensland, Brisbane, QLD, Australia
e-mail: ian.townsend@uq.net.au

S. McKinnon, M. Cook (eds.), *Disasters in Australia and New Zealand*,
https://doi.org/10.1007/978-981-15-4382-1_7

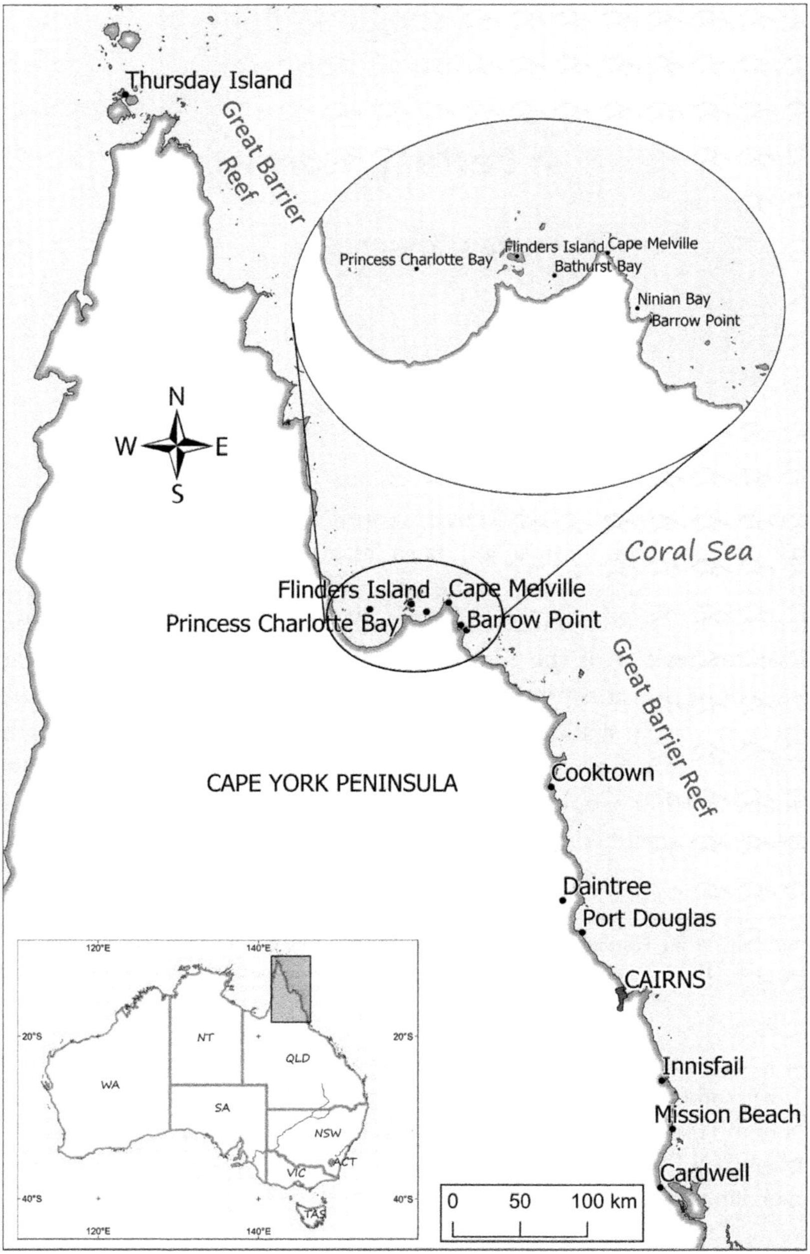

Fig. 1 Map of locations in the Cape York region, Far North Queensland

Indigenous people across northern Australia had been using the shell of the pearl oyster, *Pinctada maxima*, for ornaments and tools for centuries before Europeans began commercially exploiting pearl shell in Australia during the 1860s.[2] Technological improvements in the manufacturing of buttons and ornaments in Europe fuelled the growth in demand in the late nineteenth century and fostered a commercial pearling industry.[3] As pearl shell beds close to shore were depleted, larger, more specialised, vessels evolved to exploit beds further away from ports. Western Australia and Thursday Island become major sources for pearl shell, and in the 1870s, diving suits with air pumps allowed divers to reach shell below six fathoms and spend more time collecting. This led to a system of "floating stations," in which company-owned schooners accompanied by fleets of luggers travelled further and stayed at sea longer.[4] Most of the work was done by low-wage foreign labourers, recruited from more than 25 nations or distinct regions around the world. The majority were young men from Japan, South East Asia, and the Pacific Islands.

On 5 March 1899, the eye of a cyclone crossed Cape Melville sinking or grounding most of the 100 vessels in the region. Of the three pearling fleets consisting of 40 vessels anchored in Bathurst Bay, all but one schooner sank,[5] and the event became known as the pearling disaster of 1899. This tropical cyclone, later named *Mahina*, holds the current world record for a storm tide: 13 metres (42 feet).[6] It is considered Australia's deadliest cyclone with an official (although, as will be shown,

---

[2] John P. S. Bach, *The Pearling Industry of Australia: An Account of its Social and Economic Development* (Canberra: Department of Commerce and Agriculture, 1955), 8.

[3] John P. S. Bach, "The Political Economy of Pearlshelling," *The Economic History Review* 14, no. 1 (1961): 106.

[4] Stephen Mullins, "To Break 'The Trinity' or 'Wipe Out the Smaller Fry': The Australian Pearl Shell Convention of 1913," *Journal for Maritime Research* 7, no. 1 (2005): 219–222.

[5] The schooner *Crest of the Wave* was the only vessel to survive the eye. It was not in Bathurst Bay, but had been blown north.

[6] Jonathan Nott, Camilla Green, Ian Townsend, and Jeffrey Callaghan, "The World Record Storm Surge and the Most Intense Southern Hemisphere Tropical Cyclone: New Evidence and Modeling," *Bulletin of the American Meteorological Society* 95, no. 5 (May 2014): 757–765; World Meteorological Organization, "Tropical Cyclone: Largest Storm Surge Associated with Tropical Cyclone," *WMO World Weather and Climate Extremes Archive*, https://wmo.asu.edu/content/tropical-cyclone-largest-storm-surge-associated-tropical-cyclone [accessed 10 November 2019].

exaggerated) death toll of about 400.[7] Aside from epidemics and heat-waves, it is Australia's deadliest natural disaster, yet it is also one of its least known. Despite its high death toll, it was never considered a national tragedy and failed to become part of the nation's social memory.[8] Today, *Mahina* is barely remembered, although embedded in Cape Melville's weather-beaten landscape are the trunks of large felled trees, which still lie jammed between smaller trees in regrowth forests. The wreck of a lightship lies at the bottom of the shipping channel. There are more than 100 unmarked graves on the beaches and nearby islands.

This chapter examines the 1899 disaster and asks two questions. First, why does Australia's deadliest disaster have such a marginal place in the nation's collective memory? Second, how have incorrect details about the event come to be accepted as scientific fact? I argue that because the cyclone has been neglected by historians, and largely forgotten by the Australian public, the lack of historical inquiry has meant that untested evidence has been accepted as fact, and will be shown in this chapter to be wrong. The cyclone was a catastrophe for the non-European pearling crews and the Aboriginal people ashore, but not a national tragedy for the anticipated White Australian nation. Only 12 of those who died were European, and the relatively few European survival stories did not fit the usual disaster tropes of heroism and stoicism in the face of adversity. Yet, the stories that inaccurately recount the event are the ones that inform current science and disaster management planning, highlighting the need for historians and historical inquiry to play a greater role in disaster science.

Remembering *Mahina*

When I lived in North Queensland in the 1990s, preparing for cyclone season was an annual ritual. Local meteorologists interviewed on radio would sometimes mention the 1899 cyclone as a worst-case scenario, but

[7] Australian Institute for Disaster Resilience, "Cyclone Mahina 1899," *Disaster Resilience Knowledge Hub*, https://knowledge.aidr.org.au/resources/cyclone-cyclone-mahina-cape-york-queensland [accessed 10 November 2019]; Guinness World Records Limited, "Highest Storm Surge," in *Guinness World Records 2014* (New York: Bantam, 2014), 107.

[8] Brad West and Philip L. Smith, "Natural Disasters and National Identity: Time, Space and Mythology," *Australian and New Zealand Journal of Sociology* 33, no. 2 (August 1997): 205–215.

the stories about porpoises found high on cliffs and stones embedded in trees were both sensational and vague. Because it was not well remembered and had all the elements for a good story, I researched the *Mahina* disaster for a novel, published in 2008.[9] *The Devil's Eye* was a dramatic recreation of the event, fiction based on fact,[10] but in researching the book, it became clear that many "facts" about this cyclone did not reflect the best evidence. Scientists studying cyclones used data from this event, but relied almost entirely on previously published papers or secondary sources. No one checked the facts.

Most of the information in official databases and scientific journals today can be traced back to three sources. The main source is a memorial booklet published in 1899 by the Outridge Printing Company, owned by a Brisbane family that part-owned one of the fleets and lost two members in the disaster. *The Pearling Disaster, 1899: A Memorial* (hereafter referred to as the Outridge booklet) was considered a "veritable mine of information" by Australian meteorologist Herbert E. Whittingham, who in 1958 published the second main source, an article in a scientific journal reconstructing the event.[11] The third source is a dramatic recreation of the disaster in a 1971 non-fiction anthology entitled *Cyclone*, by Queensland journalist Hector Holthouse.[12] Much of the information published by Whittingham and Holthouse is derivative, taken from the anonymously written Outridge booklet, and they circulate the same misinformation.

Forgetting, Race, and the Nation

Ask most people today to name the worst or deadliest cyclone in recorded Australian history and they will probably say cyclone *Tracy*, which struck Darwin in 1974. Although the memory of Queensland's 1899 cyclone does persist within some communities and families, it remains largely

[9] There are some elements to a "good story" on which writers and journalists generally agree. It often includes conflict, drama, something new or surprising, and it must be relevant to an audience.

[10] Ian Townsend, *The Devil's Eye* (Pymble, NSW: HarperCollins, 2008).

[11] H. E. Whittingham, "The Bathurst Bay Hurricane and Associated Storm Surge," *Australian Meteorological Magazine* 23 (1958): 4–36.

[12] Hector Holthouse, *Cyclone* (Adelaide: Rigby, 1971), 4–15.

outside wider national discourses of notable disasters.[13] The disaster was framed to be forgotten.

In 1899, the Outridge booklet intended to provide an account "of the Europeans who were lost," compiled "from press and official reports, and from the testimony of survivors," and made available to "the immediate relatives of the deceased and to others interested."[14] It referred to the European community, which represented just four per cent of those killed in the fleets. The memorial booklet was not produced to include narratives of the foreign crews or the Aboriginal people ashore, except where they intersected with the European narratives. The booklet is a piece of journalism, and as Hayden White warns, the stories journalists tell should not be confused with historical narratives, as they are "locked within the confines" of a contemporary chronicle.[15] The Outridge booklet is a reflection of its own intentions, a cultural narrative that lacks "secondary referentiality."[16] The booklet and newspapers of the time reflected nineteenth-century social and racial discourses and, as such, are responsible for much of what was to be forgotten and remembered about the pearling disaster.

The pearling disaster, on 4 and 5 March 1899, occurred during heightened political discourses about race. Queensland Legislative Assembly elections had started on 1 March and were to continue across the colony until 25 March 1899. The Australian colonies were also in the midst of a series of referenda on Federation, with the Queensland referendum to take place in September of that year.[17] A unifying force for Federation was the desire to create a White Australia, but for the white pearling fleet owners, foreign labour was an economic issue and it posed a dilemma. On the one hand, a consortium of businessmen led by fleet owner James

[13] See, for example, Australian Government, "Historical Impacts Along the East Coast," Australian Bureau of Meteorology, http://www.bom.gov.au/cyclone/history/eastern.shtml [accessed 10 November 2019].

[14] Anonymous, *The Pearling Disaster, 1899: A Memorial* (Brisbane: Outridge Printing Company, 1899).

[15] Hayden White, *The Content of the Form: Narrative Discourse and Historical Representation* (Baltimore: Johns Hopkins University Press, 1987), 172.

[16] Ibid.

[17] Voting in the 1899 Queensland election was held at different times for different places between 1 and 25 March. The constitutional referendum in Queensland was held on 2 September 1899.

Clark had been pushing hard to stop competition from the "coloured aliens," especially the Japanese, in the northern pearl shell industry.[18] On the other hand, the industry could not operate as profitably without cheap foreign labour. It was inevitable that the pearling disaster would become conflated with the White Australia narrative, as happened when the *Brisbane Courier* published its editorial opinion on 20 March 1899:

> Those who would have diving carried on wholly by white men may now perhaps see what an awful blow would have been dealt to Queensland had four hundred families in our midst been sorrowing for loved ones lost.[19]

In 1899, an Australian nation was not yet real, but anticipated, and it can be considered within the theoretic framework described by Benedict Anderson as an "imagined community" whose members conceive of themselves as being part of the nation "regardless of actual inequality and exploitation that may prevail in each."[20] The members of the imagined Australian nation, though, were to be European and mainly British. "Coloured aliens" were to be excluded.

The deaths of about 300 non-European people in the pearling disaster did not resonate for a White Australia, and this was reflected in commemorations. A memorial at the All Souls St Bartholomew Quetta Memorial Church on Thursday Island names seven Europeans who drowned, but not of the "295 coloured men" who also died.[21] A memorial at Bathurst Bay names 12 Europeans, but none of "over 300 Coloured Men Drowned."[22] The names of most of the "coloured men" were known at the time, but appear not to have been considered worthy of com-

---

[18] Regina Ganter, *The Pearl-Shellers of Torres Strait: Resource Use, Development and Decline 1860s–1960s* (Melbourne: Melbourne University Press, 1994), 106; Stephen Mullins, "From TI to Dobo: The 1905 Departure of the Torres Strait Pearl Shelling Fleets to Aru, Netherlands East Indies," *The Great Circle: Journal of the Australian Association for Maritime History* 19, no. 1 (1997), 33; Robert Lehane, *The Pearl King* (Brisbane: Boolarong Press, 2014), 141.

[19] "Further Light on the Hurricane," *Brisbane Courier*, 20 March 1899, p. 4. Estimates of the death toll fluctuated before a list of the dead was compiled by the fleet owners and the Queensland Government.

[20] Benedict R. Anderson, *Imagined Communities: Reflections on the Origin and Spread of Nationalism* (London: Verso, 2006), 6–7.

[21] Text of the memorial plaque on Thursday Island.

[22] Text of the memorial monument at Bathurst Bay.

memoration. The 1899 cyclone was immediately framed as a disaster in which most of the victims were clearly not to be part of an imagined Australian nation.

In their study of disasters and national identity, Brad West and Philip Smith argue that cyclones, floods, and bushfires are considered transitory threats. Droughts, with their long duration and larger geographical scale, are more often considered national disasters. Cyclones, they argue, can only be constructed "with difficulty as a national event."[23] However, West and Smith see cyclone *Tracy* as one exception.[24] The victims of *Tracy* were predominantly white and considered without question to be Australian. Military-controlled relief efforts also framed the disaster as a quasi-military event, providing a link to the ANZAC myth, an existing core element of national identity since World War I.[25] The event attracted national newspaper coverage for at least two weeks and captivated the nation. The timing of the 1899 cyclone, on the cusp of Federation, meant the disaster had the potential to help define the new nation, but it failed to do so because foreign labourers dying in the far away and dangerous tropics did not reflect the narrative a White Australia was looking for. Newspapers linked the event not to the emerging new Australian nation, but to ideals of its colonial British past, as expressed in a letter from a London correspondent published in the *Week* newspaper:

> The blow is a heavy one for a small and young colony, but Queensland has, I hope, the mettle to meet it. Troubles and difficulties have often proved the grit of the British colonist. So may it be with you. In the meantime be assured that you have the warm sympathy of 'the old folks at home.'[26]

In the absence of a federal government, the disaster was considered a colonial Queensland responsibility. The Queensland Government did martial resources to help the shipwrecked sailors and the Aboriginal

[23] West and Smith, "Natural Disasters and National Identity," 208.

[24] Ibid.

[25] Brad West, "Mythologising a Natural Disaster in Post-industrial Australia: The Incorporation of Cyclone Tracy Within Australian National Identity," *Journal of Australian Studies* 24, no. 66 (2000): 200.

[26] "Our London Letter," *Week*, 28 April 1899, p. 23.

people ashore.[27] Significantly, though, a public appeal for funds for the families of dead sailors secured only three donations.[28] There was little or no newspaper coverage of the effect the cyclone had on the families of those killed.

Nor was there much mention of heroes, another aspect of disaster that can garner public sympathy and make it memorable to a community. Historian Paula Hamilton, in her study of the 1938 *Rodney* ferry disaster in Sydney, discusses conventional narrative frameworks for disasters in media coverage that concern bravery as well as "victims of fate."[29] In the 1899 pearling disaster, however, most of the Europeans who might have filled those narrative roles died when their vessels sank in Bathurst Bay.[30] The stories of heroism appearing in the press related almost entirely to the non-European crews and local Aboriginal people.

> We read of Japanese women who swam to land not knowing that the infants in their arms, for whom they struggled so heroically, were dead.[31]

> A colored man reached the shore accompanied by two colored women after swimming about four days.[32]

> Considerable bravery was displayed by the coastal aboriginals in rendering assistance to the wrecked pearlers and in burying the bodies which were washed ashore.[33]

> The natives were burying the dead and fighting the bushmen, and keeping them back from the bodies.[34]

---

[27] Walter Roth to Under Secretary Home Office, 9 April 1899, HOM/A23/99/5252, ID847561, 1, "Report Re Distribution of Gifts to Coastal Aboriginals," Queensland State Archives, Brisbane.

[28] "Pearling Disaster. Mayor of Brisbane's Fund," *Telegraph* (Brisbane), 8 April 1899, p. 9. Only three members of the public subscribed.

[29] Paula Hamilton, "Memory Remains: Ferry Disaster, Sydney, 1938," *History Workshop Journal 1999*, no. 47 (1999): 203.

[30] William Field Porter, the master of the schooner *Crest of the Wave*, provided the only narrative of the white male hero, saving his wife and child by forcing his crew to the pumps.

[31] *Sydney Morning Herald*, 17 March 1899, p. 4.

[32] "The Hurricane. A Colored Man's Bravery," *North Queensland Register*, 20 March 1899, p. 14.

[33] "Northern Hurricane," *Telegraph*, 20 April 1899, p. 2.

[34] "Return of the Warrego," *Queenslander*, 25 March 1899, p. 545.

The confusing nature of some of the reporting helped weaken a narrative, which included rumours that were later disputed.

> Mr. James Clark … spoke in high terms of the attention of the coastal aboriginals, and ridicules the idea of the hill tribes fighting the coast ones over the plunder.[35]

Sociologist Thomas R. Forrest identifies disasters as major disruptions to society, representing "one of life's most significant events, both individually and for the community. The anniversary of a disaster becomes an expression of community memory."[36] In the pearling disaster, however, there were three main communities identified: Europeans, non-European crews, and Aboriginal people ashore. Most of the surviving non-European crews were repatriated or emigrated from Australia after Federation, taking with them their cultural memory. For the Aboriginal people, the catastrophic population decrease as a result of contact with the fishing industry, the later forced removal of children from their families, and the removal of the surviving adults to separate missions in the early part of the twentieth century, severed access to many memories, although oral histories do still exist. No one lives permanently at Cape Melville anymore. The country has been, as Traditional Owner Daniel Gordon describes, "cleaned out."[37]

One of the features that distinguish the pearling disaster from other Australian disasters is the lack of commemoration in the years following the event. One small notice appeared in the *Brisbane Courier* on the first anniversary, in 1900, which read, "In memory of the Pearlers and Crew of the Lightship, drowned off Cape Melville in the hurricane of the 4th and 5th March, 1899."[38] Other newspaper memorials followed for three years only, remembering the European pearlers Alfred and Harold

---

[35] Ibid.

[36] Thomas R. Forrest, "Disaster Anniversary: A Social Reconstruction of Time," *Sociological Inquiry* 63, no. 4 (1993): 447.

[37] Daniel Gordon interviewed 25 September 2018.

[38] "Family Notices," *Brisbane Courier*, 5 March 1900, p. 4.

Outridge, Robert Cameron, and William Powell.[39] Cultural conventions normally provide a framework in which anniversaries are observed. There is often a first anniversary followed by remembrance at 5 years, and then often 25, 50, 75, and 100 years.[40] Theories of disaster commemoration fail to explain the anomaly of the pearling disaster that has largely been overlooked. Most of those drowned were considered external to the colonial society in which the disaster occurred. Cape Melville was also so remote as to be outside the experience of most Australians, who lived south of the Tropic of Capricorn. It would be decades before journalists rediscovered and reprised the event. This fragmented retelling influenced the way the disaster was to be remembered and studied.

Paula Hamilton explains that "disasters are not usually regarded as the substance of scholarly work by historians; it is left rather to other storytellers to keep these tales alive in the cultural imagination."[41] According to Hayden White, a historical event cannot be separated from its retelling, and the retelling is central to how the past is remembered.[42] In the case of the pearling disaster, it was left to journalists to later retell the event as a popular, narrative non-fiction history, and they struggled to find a narrative that resonated.

The first significant narrative retelling of the 1899 disaster occurred 35 years later, after a severe cyclone struck the coast north of Port Douglas on 11 March 1934.[43] It was rarely mentioned again until 1971, when Queensland journalist and author Hector Holthouse reprised the information from the Outridge booklet for his book *Cyclone*. The Outridge booklet had repeated the claim by Queensland Government Meteorologist, Clement Wragge, that the disaster was caused by two cyclones, *Nachon* and *Mahina*, colliding over Bathurst Bay. This was a result of Wragge revising his weather charts after the event to explain why he had failed to

[39] "Family Notices," *Telegraph*, 5 March 1901, p. 4; "Family Notices," *Queenslander*, 8 March 1902, p. 504; "Family Notices," *Brisbane Courier*, 4 March 1902, p. 4; "Family Notices," *Sydney Morning Herald*, 5 March 1903, p. 6.

[40] Forrest, "Disaster Anniversary," 446.

[41] Hamilton, "Memory Remains," 197.

[42] Hayden White, "The Historical Text as Literary Artifact," *Clio: An Interdisciplinary Journal of Literature, History, and the Philosophy of History* 3 (1974): 277–303.

[43] "1899 Holocaust Recalled," *Telegraph*, 17 March 1934, p. 7.

predict the disaster. In Wragge's original charts, *Mahina* never crossed the coast. Holthouse quite rightly dismissed Wragge's two-cyclone theory, but then changed the narrative again by declaring the disaster to be "*Mahina*'s responsibility alone."[44] The name *Mahina* is now firmly linked to the cyclone, but that is relatively recent and another example of how memories and data change over time.

Data, Data Everywhere

There are now thousands of pieces of published data on the 1899 cyclone, most in scientific literature, with little support from the earliest and best evidence. For example, many official sources today report a death toll of 400,[45] but contemporary reports from 1899 show 100 fewer deaths.[46] The central air pressure of the cyclone is listed in databases as 914 hectopascals (hereafter written as 914 hPa, originally recorded as 27 inches of mercury, or 27 inHg). However, the day after the cyclone, the captain of the schooner *Crest of the Wave* reported a low pressure of 26 inHg (880 hPa).[47] This pressure is so low that it would today constitute the lowest recorded pressure of a cyclone in the Southern Hemisphere. Depending on the scientific paper cited, the storm tide associated with the cyclone is between 12 and 14.6 metres, and yet all the measurements stem from a single observation cited in the Outridge booklet. An officer of the Native Police, Constable John Martin Kenny, coincidentally ashore at the time, described the sea coming up to his waist while standing on a ridge "fully 40 feet" (12.19 metres) above sea level.

---

[44] Holthouse *Cyclone*, p. 15. One newspaper article in 1955, however, did link it to *Mahina* before 1971. See "Looking at Queensland," *Beaudesert Times*, 30 December 1955, p. 6.

[45] Australian Institute for Disaster Resilience, "Cyclone Mahina 1899."

[46] Anonymous 1899. The Outridge booklet reports 307 deaths. See also George Bennett, "Return Giving Names and Nationality of Persons Belonging to Pearling Fleet Lost in Hurricane of 4th and 5th March, 1899, in Neighbourhood of Cape Melville," in T. M. Almond, "Report on the Marine Department for the Year 1899–1900," *Queensland Legislative Assembly Votes and Proceedings* 3 (24 September 1899): 24. Bennett reports 247 names.

[47] William Field Porter, "The Great Hurricane at Queensland. A Struggle for Life. An Aucklander and His Wife and Child. Forcing the Blacks to the Pumps," *New Zealand Herald* (Supplement), 1 April 1899, p. 1.

Despite its extraordinary severity and high death toll, the 1899 cyclone had been left out of many lists of Australian disasters. For example, in examining climate in relation to culture, the National Museum of Australia's 2005 publication, *A Change in the Weather: Climate and Culture in Australia*, did not mention cyclone *Mahina*, but described cyclone *Tracy* as "the worst storm in recorded Australian history."[48] In the Australian Bureau of Meteorology's selected list of historic impacts along the east coast of Queensland, the 1899 cyclone is missing, while other major impacts are described.[49] The deep and long-standing separation of our deadliest cyclone from our cultural narratives persists, and this is a potentially dangerous oversight.

Environmental risk scientist Ken Granger argues that a community's stories about previous disasters are vital for disaster management because they help "overcome the inherent problem that human memory tends to be significantly shorter than the return period of most hazard phenomena."[50] Remembering a disaster accurately is literally a matter of life and death. History, like science, is a discipline and scholarly historians play a vital role in disaster research by checking the facts. In the case of one of Australia's deadliest disasters, though, scholarly historians and historiography are missing. As Arthur Marwick says, historian's skills "lie in sorting these matters out, in understanding how and why a particular source came into existence, [and] how relevant it is to the topic under investigation."[51] Their skills also lie in testing the evidence, including data, for reliability.

"What do historians do? I have no idea!" one scientist, whom I greatly respect, asked me during my research. Most scientists never work with historians. Even if studying the same event, such as a disaster, natural scientists and social scientists are usually interested in different aspects.

---

[48] Bill Bunbury, "Cyclone Tracy: Voices on the Wind," in *A Change in the Weather: Climate and Culture in Australia*, ed. Tim Sherratt, Tom Griffiths, and Libby Rodin (Canberra: National Museum of Australia Press, 2005), 165–173. Bunbury examines cyclone *Tracy*, "the worst storm in recorded Australian history."

[49] "Historical Impacts Along the East Coast," Australian Bureau of Meteorology.

[50] Ken Granger, "An Information Infrastructure for Disaster Management in Pacific Island Countries," *Australian Journal of Emergency Management* 15, no. 1 (Autumn 2000): 25.

[51] Arthur Marwick, *The New Nature of History: Knowledge, Evidence, Language* (Basingstoke, UK: Palgrave, 2001), 27.

What they both do is answer research questions and solve problems. Scientists and historians try to reach a hypothesis based on probability. The better the evidence, the greater the level of confidence in their hypothesis. What actually happened? What was the height of the storm tide? How many people died? Why is the event so poorly remembered? What does that mean for coastal communities? In answering these questions, both historians and scientists try to get as close to the truth as possible.

Historians debate representations of history, aware that the complex interplay between culture and memory can distort evidence and historical accuracy. Scientists draw on historical accounts for data, but if they do not understand what historians do, they may not be using the best data. This case study not only highlights how historical accounts of cyclone *Mahina* and the pearling disaster have been distorted, but it raises serious concerns about how they have been interpreted by science and disaster managers to prepare for future events.[52]

The Death Toll

One of the significant pieces of data cited by scientists is the 1899 cyclone's death toll, often stated to be 400.[53] This figure is said to include 100 Aboriginal people ashore swept into the sea, but no reference to that figure appeared until the 1970s. It could reasonably be expected that such a significant part of the death toll would be mentioned at the time, or before 1971, if there was an earlier source.[54] The President of the Royal Historical Society of Queensland, Norman Pixley, for example, did not mention it when he delivered his presidential address about the pearling disaster at the society's annual meeting on 23 September 1971.[55] However, after 1971, reports that 100 or "about 100" Aboriginal people were swept

[52] Granger 2000, 25.

[53] Australian Institute for Disaster Resilience, "Cyclone Mahina 1899"; Guinness World Records Limited, "Highest Storm Surge," 107.

[54] Only eight deaths were recorded ashore by Walter Roth, during a survey after the cyclone. It could be speculated that there were other deaths ashore, but there is no evidence.

[55] Norman S. Pixley, "Presidential Address: Pearlers of North Australia: The Romantic Story of the Diving Fleets," *Journal of the Royal Historical Society of Queensland* 9, no. 3 (1972): 9–29.

out to sea either by a tidal wave or by a wind change began to appear in articles and books. That reference to 100 Aboriginal people drowning first appears in Hector Holthouse's popular narrative non-fiction book *Cyclone*, first published in 1971.

> A final count of those who lost their lives was 307: twelve white men and 295 Asiatics and Islanders. This did not take into account about 100 Aborigines swept out to sea and others killed in the forest country.[56]

In his book, Holthouse links these deaths to the world record storm tide, writing that "the back surge from the wave which caught Constable Kenny" swept the 100 Aboriginal people away. There is no evidence for this. The first iteration of 100 Aboriginal people being swept out to sea and drowning appears in the corrections to the first draft of Holthouse's manuscript, held at the John Oxley Library.[57] The words "a considerable number of" are crossed out and replaced with "about 100." Holthouse appears to have invented the 100 Aboriginal dead, but the figure found its way into a government database and, in 2012, a researcher at the Australian Maritime College used it as an example of the sort of data that could be fed into mathematical models to predict "storm surge risk under current and future climate conditions." The article describes how the 100 people may have been swept into the sea.

> Many of these aborigines were drowned because *Mahina* blew in from the east and skimmed along the northward facing beaches of Princess Charlotte Bay. ... The population of synthetic cyclones can help model unexpected situations like this. No one could have thought Cyclone *Mahina* would have the outcome that it did.[58]

The Flinders Island, Bathurst Bay, and Barrow Point people, and the Guugu Yimithirr-speaking people south-east of Princess Charlotte Bay, are the traditional owners of the region. Yiidhuwarra elder Daniel Gordon, a

---

[56] Holthouse, *Cyclone*, 13.

[57] Hector Holthouse, *Cyclone* manuscript draft, n. d., M797, 34, Hector Holthouse scrapbooks, manuscripts and cuttings, John Oxley Library, State Library of Queensland, Brisbane.

[58] "Tropical Cyclone Wave Modelling," *Shore to Sea 1*, Australian Maritime College, University of Tasmania (July 2012): 37.

Traditional Owner of Flinders Island and a manager and ranger of the Cape Melville National Park that covers these areas, is the adopted son of Bob Flinders, who passed the story of the 1899 cyclone down to him. Daniel Gordon said some of the Flinders Island people were visiting Cape Melville at the time of the cyclone. They knew a cyclone was coming, so they sheltered in a cave in the Cape Melville Range. As Daniel Gordon explained to me, "the people from around Cape Melville area, Bathurst Bay, they were taking notice of the tides and they knew there was something wrong with the weather because the tide was unusual … So they, you know, they went through all this during the past, so they moved it into a higher ground. There's a shelter at Cape Melville where they all moved to."[59]

After the cyclone, Aboriginal people helped find and bury the dead pearlers, most of whom, Daniel Gordon said, were "Islanders and a few, I think, Aboriginals were on board, too. Most of them I think Islanders as well." I asked him whether he knew of a story about 100 local people who were swept out to sea. "No, it was only the people who were working on the luggers."[60]

The Implications of Forgetting

The Queensland *Disaster Management Act 2003* guideline on mitigating the impacts of cyclones considers "worst case" scenario predictions for storm tides.[61] One problem in preparing for cyclones is that no one really knows what a worst-case scenario looks like. Category Five cyclones passing directly over populated areas and producing damaging storm tides have been so infrequently recorded that it is difficult to get a statistically significant sample.[62]

---

[59] Daniel Gordon interviewed 2 March 2015.

[60] Ibid.

[61] Queensland Government, *Queensland Prevention Preparedness, Response and Recovery Disaster Management Guideline* (Brisbane: Queensland Fire and Emergency Services, 2017), 27, https://www.disaster.qld.gov.au/dmg/Documents/QLD-Disaster-Management-Guideline.pdf. Government risk assessments include the use of historical analysis "to cover the spectrum of most likely to credible worst case scenarios using geospatial intelligence."

[62] Bruce Harper, "Storm Tide Threat in Queensland: History, Prediction and Relative Risks," *Conservation Technical Report No. 10*, RE 208 (Canberra: Department of Environment and Heritage, 1999), 5. "The historical rarity of severe storm tides along the Queensland coast, and

The 1899 cyclone was unique. Not only was it the most severe recorded on landfall, but the damaging eye passed over and between a relatively large number of trained and reliable weather observers, including ships captains who made regular observations using thermometers, barometers, and compasses. The data from this cyclone provides an opportunity to learn from a rare and well-recorded event. The value of surviving observations to scientists and emergency managers is partly recognised in the numerous academic papers that have been published using data associated with those observations. However, much of the data they have been using to describe the event, including data appearing in the scientific literature, comes not from the surviving earliest primary sources but from later reports, secondary sources, rumour, and memory. Memory is useful to historians, to describe the social aspects of the disaster, but science requires a quantitative approach. Scientists need reliable data, particularly measurements, to help us all understand cyclone behaviour.

The current entry for the 1899 cyclone in the Australian Government's database on disasters, the *Disaster Resilience Knowledge Hub*, is a compilation of data from untested secondary sources:

> On 4 March 1899, a Category Five cyclone hit Bathurst Bay with winds reaching 260 kilometres per hour. A tsunami of 14.6 metres swept inland for five kilometres. A wave surge measuring 13 metres at Ninian Bay (adjacent to Barrow Point, 30 kilometres south of Bathurst Bay) extended inland for three to five kilometres. At Bathurst Bay, near Princess Charlotte Bay (Cape York), at least 307 crew members died from a pearling fleet of over 100 vessels plus other craft (with 152 sunk or wrecked, some found kilometres inland) as a result of the storm surge. Over 100 Aboriginal people died trying to help the shipwrecked.[63]

---

more particularly at any specific location, precludes the use of basic statistical methods using measured water level data for the estimation of long term risk." Bruce Harper, Stan Stroud, Michael McCormack, and Steve West, "A Review of Historical Tropical Cyclone Intensity in Northwestern Australia and Implications for Climate Change Trend Analysis," *Australian Meteorological Magazine* 57 (2008): 121–141. The study notes that the eyes of fewer than 20 of 900 cyclones in the Australian database passed directly over an instrument that could reliably measure them.

[63] Australian Institute for Disaster Resilience, "Cyclone Mahina 1899."

In fact, the cyclone crossed the coast on 5 March, the winds were not measured, and there was no tsunami. A historical methodology applied to the historical documents shows that there is a reliable report of a storm tide that may be as high as 13 metres, but that height was not measured and is yet to be corroborated. It was not reported at Ninian Bay. Although the death toll may have exceeded 300, there is evidence for fewer than 300 deaths.[64] Fewer than 60 vessels were totally lost (some vessels were raised and salvaged), and none was found kilometres inland. There is no evidence that 100 Aboriginal people died, but we do not know the true figure.

Credible primary source data on the 1899 cyclone have always been available, but media errors and social and political narratives over 120 years have distorted the historical record, government databases, and the scientific literature. There has been relatively little interest in Australia's historical cyclones by historians, and so the historical inquiry has been left largely to disaster scientists.

Disaster science is booming. Journal publisher Elsevier, in its *A Global Outlook on Disaster Science* report, noted that 27,000 papers on disaster science had been published between 2012 and 2016, many of them about prevention and preparedness, stressing the need for good data to inform policymaking. Disasters are becoming more costly. In 2017, cyclone *Debbie* in Queensland cost the insurance industry $1.67 billion, creating a substantial financial incentive to reduce the risks and to better understand the impacts of climate change on cyclone behaviour and how cyclones cause damage. In this area of science, scientists rely on history to model the future, and so it is important to get the history right. It is worth considering whether other historical cyclones are as factually flawed as our understanding of the 1899 cyclone. Any scientific evaluation of past cyclones, the data for which is drawn largely from historical records, should include a review of the evidence for that data, preferably by historians. The lives of thousands of people in communities along Australia's northern coasts will depend on it.

[64] Ian Townsend, "The Bathurst Bay Hurricane: Media, Memory and Disaster" (PhD thesis, University of Queensland, 2019).

Part III

Legacies of the Past

Shallow Fire Literacy Hinders Robust Fire Policy: Black Saturday and Prescribed Burning Debates

Daniel May

On Saturday, 7 February 2009, the state of Victoria experienced its worst ever disaster. A severe heatwave, record-breaking temperatures and intense winds caused a series of bushfires to grow into firestorms, killing 173 people, destroying thousands of houses and burning nearly half a million hectares (Fig. 1).[1] In the aftermath of the disaster, public controversy began to rage in the public sphere and to the Royal Commission entrusted with investigating the fires. Debate particularly focused on the role of prescribed burning and how greater use of it might have prevented the tragedy. In this chapter I explore how this example of settler Australians reacting to a disaster demonstrates how the nation's non-Indigenous people are still grappling with fire. The popular debate and 'blame game' over prescribed burning reflected both a lack of fire literacy and the

[1] Victoria. Bushfires Royal Commission et al., "2009 Victorian Bushfires Royal Commission, Final Report: Summary" (Melbourne: Government Printer for the State of Victoria, 2010).

D. May (✉)
School of History, Australian National University, Canberra, ACT, Australia
e-mail: daniel.may@anu.edu.au

S. McKinnon, M. Cook (eds.), *Disasters in Australia and New Zealand*,
https://doi.org/10.1007/978-981-15-4382-1_8

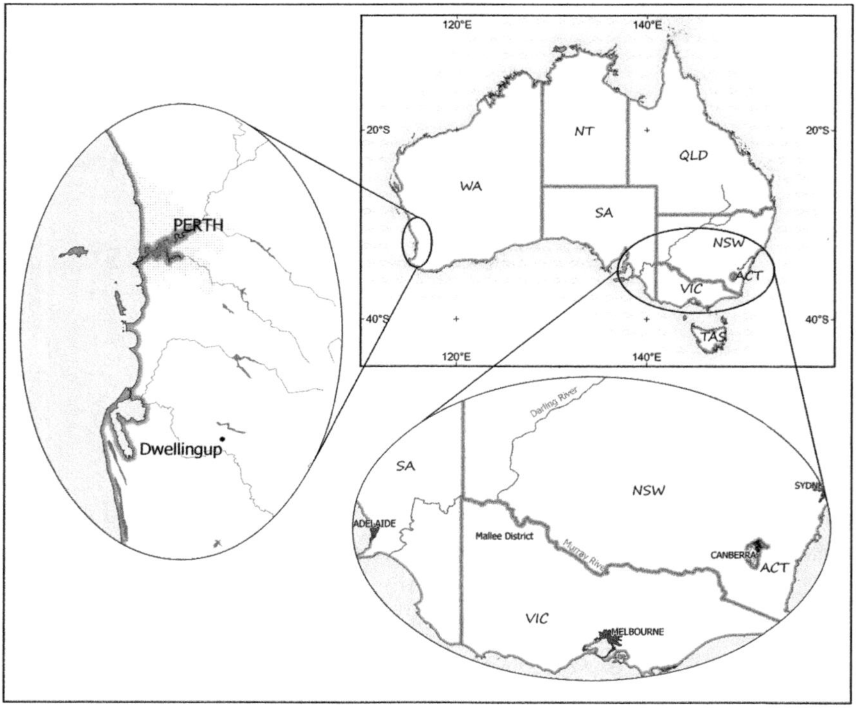

Fig. 1 Map of key locations in Victoria and Western Australia

appropriation of fire into broader culture wars. The policy response to Black Saturday—a simple hectare-based target for prescribed burning—lasted just five years and was a failure; it did not reach the nominated performance measures and it was too ecologically blunt and inattentive to ecological nuance. The debates and policy shifts after Black Saturday illustrate that Australians cannot reach more effective fire practice without better fire policy and cannot legitimate better fire policy without deeper fire literacy.

Prescribed burning is known globally by many terms, including 'controlled burning' and 'planned burning'. It can be defined as "the controlled application of fire under specified environmental conditions to a predetermined area and at the time, intensity and rate of spread required to attain planned resource management objectives".[2] Prescribed burning can be conducted to reduce the fuel available for future bushfires (often known as 'hazard reduction burning'), for ecological management purposes (such as to stimulate the growth or propagation of certain species, perhaps for conservation or to provide increased grazing fodder), to mitigate against invasive species, or for cultural reasons.[3] While there is always the risk of prescribed burns escaping, it can be a strongly effective component of a successful fire management system in multiple ways. For example, it may reduce the amount of flying embers which can ignite houses,[4] or it may slow the spread of a bushfire, increasing the chances of suppression.[5] The complexities of prescribed burning require nuanced discussion—a feature sadly lacking from public discussion after Black Saturday, and reflected poorly in the policy response.

The History of Prescribed Burning Politics in Australia

Characterised by American fire historian Stephen Pyne as "the fire flume" due to its unique combination of climate and winds, there is a long history of devastating bushfires in Victoria, which have ignited many debates and policy responses focussed on prescribed burning.[6] Indeed, Victorian fire history is dominated by the shadow cast by Judge Leonard Stretton and his three influential Royal Commissions on land management. One enduring influence of Stretton's Commission into the 1939 Black Friday

---

[2] Australasian Fire and Emergency Service Authorities Council, "Bushfire Glossary" (Melbourne: AFAC, 2012).

[3] Australasian Fire and Emergency Service Authorities Council and Forest Fire Management Group, "Overview of Prescribed Burning in Australasia," Report for the National Burning Project—Subproject 1, 2015.

[4] Philip Gibbons et al., "Land Management Practices Associated with House Loss in Wildfires," *PLOS ONE* 7, no. 1 (2012): e29212.

[5] Matthias M. Boer et al., "Long-Term Impacts of Prescribed Burning on Regional Extent and Incidence of Wildfires—Evidence from 50 Years of Active Fire Management in SW Australian Forests," *Forest Ecology and Management* 259, no. 1 (2009): 132–142.

[6] Stephen J. Pyne, *Burning Bush: A Fire History of Australia* (New York: Henry Holt and Company, 1991), 339.

bushfires has been the expectation that large Victorian bushfires will be followed by some public form of official inquiry. These inquiries (such as Royal Commissions or parliamentary committees) have helped establish "a widespread public expectation that wildfires are always preventable".[7] This particularly shapes public discussion and discourse around policies towards prescribed burning.[8]

In his 1939 Report, Stretton noted that the "amount of [prescribed burning] which was done was ridiculously inadequate" and that the Victorian Forests Commission "must recognise the necessity of protective burning in its areas"—largely referring to "strip" and "patch" burning.[9] This was the first sanctioning, from an official source, of controlled burning for fuel reduction in Australia.[10] Importantly, Stretton qualified this: "[I]t is not suggested that the practice be followed in mountain ash (*Eucalyptus regnans*) country, except to a small extent, where necessity demands that it should be done"—displaying a level of attentiveness to ecological nuance far beyond that of many voices in the Black Saturday debate 70 years later.[11] This is in essence the eternal drama of prescribed burning. As fire scientist Malcolm Gill argues, "[W]hile particular fire regimes may be more desirable than others for the wellbeing of the biota, the best fire protection of human life and property is afforded by minimal fuel quantities maintained by frequent fires of low intensity … the fire

[7] Timothy Neale, Jessica K. Weir, and Tara K. McGee, "Knowing Wildfire Risk: Scientific Interactions with Risk Mitigation Policy and Practice in Victoria, Australia," *Geoforum* 72 (June 2016): 22.

[8] See Josh Whittaker and David Mercer, "The Victorian Bushfires of 2002–03 and the Politics of Blame: A Discourse Analysis," *Australian Geographer* 35, no. 3 (September 2004): 259–287; Tom Griffiths, "How Many Trees Make a Forest? Cultural Debates about Vegetation Change in Australia," *Australian Journal of Botany* 50 (2002): 375–389; Stephen J. Pyne, *The Still-Burning Bush* (Melbourne: Scribe Short Books, 2006); Deb Anderson, Philip Chubb, and Monika Djerf-Pierre, "Fanning the Blame: Media Accountability, Climate and Crisis on the Australian 'Fire Continent'," *Environmental Communication* 12, no. 7 (2018): 928–941.

[9] Victoria. Royal Commission to Inquire into the Causes and of and Measures Taken to Prevent the Bush Fires of January 1939, "Report of the Royal Commission to Inquire into the Causes of and Measures Taken to Prevent the Bush Fires of January, 1939 and to Protect Life and Property and the Measures to Be Taken to Prevent Bush Fires in Victoria and to Protect Life and Property in the Event of Future Bush Fires" (Melbourne: Government Printer, 1939), 16, 31.

[10] Pyne, *Burning Bush*, 313.

[11] Victoria. Royal Commission into the Bush Fires of January 1939, "Stretton Commission Report," 31.

regime best suited to the biota may differ widely from that best suited to fire protection".[12]

Prescribed burning was not implemented throughout Australia in an identical fashion; one strong example of alternative practice drawn upon in the post-Black Saturday debate was represented by Western Australia (WA) where the "Australian strategy" found its fullest expression.[13] WA had led the way in experimenting with "broad-area" prescribed burning (i.e. aiming to reduce fuel not just along tactically sited strips but throughout entire forests) especially after a 1954 policy shift.[14] This was confirmed by the 1961 Rodger Royal Commission, which investigated the Dwellingup bushfires and directed the Western Australian Forests Department to "improve and extend the practice of control burning".[15] After the world's first aerial prescribed burn was conducted at Shannon River in 1965,[16] bushfire scientist Alan McArthur was able to claim in 1973 that "[i]nstead of the Aboriginal firestick, we now use aircraft dropping incendiary capsules which light up the country in a grid pattern and produce a mosaic pattern of burnt and unburnt land".[17] Over the next few decades, the Department of Conservation and Land Management was able to burn significant parts of its jurisdiction. In the Warren area, for instance, it successfully annually prescribe burnt between 5 and 10 per cent of its 900,000 hectares of largely jarrah (*Eucalyptus marginata*) and marri (*Corymbia calophylla*) open forest. This had the effect of significantly reducing the number and size of unplanned fires, though only

---

[12] A. Malcolm Gill, "Post-Settlement Fire History in Victorian Landscapes," in *Fire and the Australian Biota*, ed. A. Malcolm Gill, R. H. Groves, and I. R. Noble (Canberra: Australian Academy of Science, 1981), 93.

[13] Pyne, *Burning Bush*, 337.

[14] Neil Burrows and Lachlan McCaw, "Prescribed Burning in Southwestern Australian Forests," *Frontiers in Ecology and the Environment* 11, no. s1 (August 2013): 3.

[15] G. J. Rodger, "Report of the Royal Commission Appointed to Enquire into and Report upon the Bushfires of December 1960 and January, February and March 1961 in Western Australia. The Measures Necessary or Desirable to Prevent and Control Such Fires and to Protect Life and Property in the Future. And the Basic Requirements for an Effective State Fire Emergency Organisation" (West Australian Parliament, 1961), 58.

[16] Roger Underwood, *Fire from the Sky: A Personal Account of the Early Days of Aerial Burning in Western Australia* (Palmyra: York Gum Publishing, 2015), 14.

[17] Alan McArthur quoted in Pyne, *Burning Bush*, 340.

when prescribed burning was rotated through every six years or sooner.[18] This is relevant because, as we will see, Western Australia's apparent success at prescribed burning became pointed to as a model for Victoria.

Victorian Premier John Brumby reacted to Black Saturday quickly, establishing the 2009 Victorian Bushfires Royal Commission on 16 February. The Commission was given broad terms of reference, with scope to both investigate and make recommendations relating to bushfire preparations (which would include fuel reduction). The Commission held 155 days of hearing, received roughly 1700 submissions, heard from 434 witnesses, received more than 1000 exhibited documents into evidence and generated over 20,700 pages of transcript.[19] As with Judge Stretton's 1939 Royal Commission into the Black Friday bushfires, the 2009 Commission chose to democratise its evidence-collection process as much as possible. Early informal consultations with communities affected by the fires were held in March and April 2009, hearings were generally live-streamed over the internet, and interim papers released. These decisions, the scale of devastation from Black Saturday, and Victoria's history of public debates over fire management, meant that a vigorous public debate over fire management was ensured. While other topics debated both by the public and by experts included the culpability of individual officials, the role of climate change or the merits of the 'stay or go' policy,[20] prescribed burning and fuel levels became one of the most contentious issues. Of the estimated 1700 submissions to the Commission, prescribed burning was identified as the topic which received the most submissions (574).[21] Whether in submissions or in the press, prescribed burning inflamed passions—indicating an apparent degree of fire literacy that on closer examination was superficial.

[18] Boer et al., "Long-Term Impacts of Prescribed Burning."

[19] Victoria. Bushfires Royal Commission et al., "2009 Victorian Bushfires Royal Commission, Final Report, Volume III: Establishment and Operation of the Commission" (Melbourne: Government Printer for the State of Victoria, 2010), 2.

[20] See, for example, Tom Griffiths, "We Have Still Not Lived Long Enough," *Inside Story*, 16 February 2009, http://insidestory.org.au/we-have-still-not-lived-long-enough/.

[21] Victoria. Bushfires Royal Commission et al., "Final Report, Volume III," 6.

Popular Debate: Not Discussing Fire on Fire's Terms

The tone of debate was established by conservative commentator Miranda Devine who argued in *The Sydney Morning Herald* the week after the fires, "It is not arsonists who should be hanging from lamp-posts but greenies".[22] According to Devine, environmentalists had prevented fuel reduction from being conducted both by direct action and through the supposed infiltration of environmentalist ideology into government agencies. The hostile atmosphere revealed and encouraged by Devine's charged criticism continued in submissions. For example, businessman Ray Evans (once named as a member of the 'Greenhouse Mafia') argued that environmentalism as represented by radical activist Paul Watson of Sea Shepherd was now effectively the "established religion" of Australia, that "hardline greens" who believed trees to be sacred had infiltrated the "Department of Scorched Earth", and that the insidious regulatory choking of the native timber logging industry by these supposed infiltrators and their allies in shire councils had led to the tragedy on Black Saturday.[23] Conservative Federal MP Wilson Tuckey largely agreed, arguing that parliamentarians had ignored past inquiries supporting prescribed burning in order to harvest environmentalist preferences in elections.[24] Even some of the more reasoned criticisms of environmentalists suggested duplicity. Max Rheese of the Victorian Lands Alliance believed that environmentalists tactically positioned themselves as offering qualified support for prescribed burning while actually being opposed to it, that grudging support was given only with crippling qualification and that environmentalists suffered from a misconception that there was a lack of evidence showing the effectiveness of prescribed burning.[25]

---

[22] Miranda Devine, "Green Ideas Must Take Blame for Deaths," *The Sydney Morning Herald*, 12 February 2009.

[23] Ray Evans, *Submission to the Royal Commission into the February 2009 Victorian Bushfires*, May 2009, http://bit.ly/2mrdGJn.

[24] Wilson Tuckey, *Submission by the Hon Wilson Tuckey MP for O'Connor To the 2009 Victorian Bushfires Royal Commission*, April 2009, http://bit.ly/2mrdGJn.

[25] Max Rheese (Victorian Lands Alliance), *Submission to the Bushfires Royal Commission*, May 2009, http://bit.ly/2mrdGJn.

Not all supporters of prescribed burning were so florid in their advocacy, though even more moderate views were viewed with inevitable suspicion by environmentalist groups. The Mountain Cattlemen's Association of Victoria, a lobby group of graziers from the Victorian High Country, argued that allowing cattle back into national parks would reduce the intensity of blazes, enabling a form of fuel reduction through grazing.[26] The Victorian and National Associations of Forest Industries suggested that the forestry industry had a great deal of expertise around fire and could play a greater role in bushfire prevention through engaging in more prescribed burns and reducing fuel through "thinning".[27]

Activist responses to hyperbolic and measured criticism are revealing of both concern around the motivation of prescribed burning advocates, and how these concerns hindered any opportunity for responding to fire on fire's terms. Greens Senator Bob Brown claimed "Greens' policy supports the ecologically appropriate use of fire", while The Wilderness Society argued it was "inappropriate, opportunistic and grossly insensitive" for anyone "from the anti-parks, pro-logging lobby to push their agenda" while the fires continued.[28] Indeed, the legacy of suspicion left by decades of Forest Wars meant any proposals for greater prescribed burning delivered by the forestry and logging industries were regarded with great suspicion by environmentalists. Simon Birrell of the Otway Ranges Environment Network accused the logging industry of a "dirty trick" campaign and argued that if anything forestry increased bushfire risks. The logging industry allegedly represented a drag on management efforts as foresters needed to conduct prescribed burns for timber regeneration, consuming personnel and resources during the short window in which burns could take place.[29] Ultimately, this deep suspicion hindered activist groups from giving a coherent response to the issue of prescribed

---

[26] Mountain Cattlemen's Association of Victoria, *Submission by the Mountain Cattlemen's Association of Victoria to the 2009 Victorian Bushfires Royal Commission*, January 2010, http://bit.ly/2mrdGJn.

[27] Victoria. Bushfires Royal Commission et al., "2009 Victorian Bushfires Royal Commission, Final Report, Volume II: Fire Preparation, Response and Recovery" (Melbourne: Government Printer for the State of Victoria, 2010), 294.

[28] Stephen Lunn, "Greenies Blamed for Fires' Scale," *The Australian*, 12 February 2009.

[29] Otway Ranges Environment Network and Melbourne Water Catchment Network, *Submissions to the 2009 Victorian Bushfires Royal Commission*, May 2009, http://bit.ly/2mrdGJn.

burning—too many battles on too many other fronts had left them wary of wolves in sheep's clothing.

The accusations levelled at environmentalists of being opposed to prescribed burning were not without some foundation; many environmental groups were hardly enthusiastic in their endorsement of it as a bushfire mitigation measure. Most wanted more evidence of its efficacy and were wary of negative effects on biodiversity or native vegetation. They were particularly sceptical of the idea of prescribed burns in wet ash forests, believing that this would harm the forest and actually promote more flammable species.[30] A report was commissioned by three prominent groups to argue that prescribed burning may have had some effect on Black Saturday but was ultimately ineffective under the extreme conditions on the day. This report reflected the growing concern environmentalist groups felt for Black Saturday as a possible harbinger of a fiery future in a warming planet and it engaged in an essentially superficial analysis of climate change to argue the fires were "unprecedented".[31] It is telling that this report was most concerned in arguing that coverage of the fires had disproportionately focussed on public rather than private land—in essence, forming a broader defence of public lands.

In their submissions to the Royal Commission, advocates for prescribed burning often failed to consider the nuances identified by environmentalists. Ray Evans quoted fire scientist Ross Bradstock's caution about the effectiveness of prescribed burning in extreme conditions at length, but inexplicably preferred to discuss the lack of property rights in the Soviet Union than engage with this rather important issue.[32] Max Rheese acknowledged that wet ash (including both *Eucalyptus regnans* and *Eucalyptus delegatensis*) forests could be difficult to prescribe burn, but dismissed other environmentalist concerns by arguing that even if prescribed burning was ineffective during extreme conditions, it does assist firefighting in more

[30] For instance, see Upper Yarra & Dandenongs Environmental Council, *Submission to the Bushfire Royal Commission*, May 2009, http://bit.ly/2mrdGJn.

[31] Richard Hughes (on behalf of Australian Conservation Foundation, Victorian National Parks Association and The Wilderness Society Victoria), *Victorian Fires February 2009: A Report on Driving Influences and Land Tenures Affected*, September 2009, http://bit.ly/2mrdGJn.

[32] Evans, *Submission.*

moderate conditions.[33] Similarly, activist groups largely failed to qualify their comments for ecological communities less vulnerable to prescribed burning or to address the operational benefits claimed from prescribed burning. These are nuanced, complex issues. The failure to discuss or even recognise this complexity is indicative both of shallow fire literacy and of the fact that these debates were as focussed on a renewed skirmish in the culture wars than the development of effective policy.

Hyperbolic arguments from hyper-partisan commentators such as Devine and Evans were directed as much at delegitimising environmentalist activist groups as at addressing policy issues. The question of whether the accusations of blame were accurate is unclear. Activist groups have held differing positions on prescribed burning which have evolved over time; the Australian Conservation Foundation, for instance, as early as 1970 published a very reasoned discussion of burning.[34] Whether their public position matched private lobbying is another consideration, as is the pragmatic question of whether activist groups even possessed a sufficient degree of influence over both policy and practice. There is also the related accusation that a more nebulous green ideology was at fault, not through direct activist action but through influencing staff members or even institutional culture of government agencies. Whether or not these accusations are true, they were certainly politically potent. Furthermore, they hindered an effective social response to bushfires by focussing political energy on blame games rather than identifying areas of consensus.

Policy Response: The Short Life of a Disaster Policy

The Commission itself chose to be guided on fuel management by a panel of experts,[35] who presented a complex view of the benefits and limitations of prescribed burning that culminated in the judgement that prescribed

---

[33] Rheese, *Submission.*

[34] Australian Conservation Foundation, *Bushfire Control and Conservation*, Viewpoint Series, No. 5 (Parkville, Victoria, 1970).

[35] The panel included fire ecologists, fire behaviour scientists and experienced land managers.

burning "is the most effective mechanism" to reduce fuel.[36] Research of previous case studies in Victoria demonstrated the factors which determine the effectiveness of prescribed burning: for instance, that weather was the most important factor in determining fire intensity and that the benefits of prescribed burning diminish over time as fuel grows back. Members of the panel noted that a major issue in fire management is the risk of 'spotting' (fires started by embers flung ahead of the main fire front) and thus 'size does matter', and that for burns to be effective in preventing spotting they need to be sufficiently large to capture the majority of falling embers (1000 hectares or greater).[37] The panel's assessment of the effectiveness of prior prescribed burning on Black Saturday itself demonstrates the issues with a simple assessment of efficacy. Previous prescribed burns could not stop the fires during the catastrophic conditions (especially during the wind change), but they may have reduced ember production and slowed the fires down. Furthermore, they had made a significant contribution to the ultimate containment of the fires (accomplished in less challenging conditions).[38]

The panel unanimously recommended that Victoria should perform more prescribed burning but gave a careful and nuanced discussion around a proposed measure of state-wide targets. Unfortunately, this was ultimately not reflected in the Commission's findings. A target, the panel agreed, should be a guide, as not every hectare burned is of equal value in reducing risk and thus a blanket target considered in isolation would not be helpful. Detailed studies were necessary that would likely show "some areas should be burned more, and that others should be burned less", based on weighing up risk reduction and the strategic value of burning against adverse ecological outcomes.[39] The expert panel took great pains to emphasise this kind of nuance; the 5 per cent target originally suggested by a 2008 Government Inquiry and heavily promoted during the

[36] Victoria. Bushfires Royal Commission et al., "Final Report, Volume II," 280.

[37] Ibid.

[38] Ibid., 282–284.

[39] Victoria. Bushfires Royal Commission et al., *Fuel Management Topic Facilitated Experts' Conference 20 February 2010: Summary of Discussion by Panel*, February 2010, http://bit.ly/2kqL3LE.

Commission was not considered by them to be a panacea.[40] The panel also heavily emphasised that the recommended increase in prescribed burning needed to be accompanied by an increased commitment to monitor and research Victoria's fire ecology, including the impact of the implementation of the policy upon landscapes. The Commission itself supported these recommendations for more research, but the panel's nuanced findings on prescribed burning became translated into the simple Recommendation 56: "The State fund and commit to implementing a long-term programme of prescribed burning based on an annual rolling target of 5 percent minimum of public land".[41] The Commission's decision to endorse this simple target also reflects its concerns around the "inexplicable" failures of Victorian agencies to provide detailed data on prescribed burning despite multiple prior inquiries recommending this.[42]

Some informed commenters and the Commission itself pointed to the success of prescribed burning programmes in south-western WA, though there was a lamentable lack of engagement with discussing the similarities and differences that might bolster or preclude this example. A group of former Western Australian fire managers pointed to WA's lack of major bushfires since 1961 as evidence of the success of their strategy, but while they acknowledged that wet ash forests would be difficult to burn under mild conditions, argued that empirical research in WA had demonstrated the effectiveness of prescribed burning in periods of high (but not catastrophic) fire danger.[43] Similarly, the Commission clearly felt the success in burning between 7 and 8 per cent of south-western land per annum demonstrated a successful and replicable model for Victoria, paying only scant attention to the ecological, topographical and weather differences that might hinder a root and branch adoption of this model.[44] Environmental historian Tom Griffiths has criticised the Commission for

---

[40] Victoria et al., *Report of the Natural Resources Committee on the Inquiry into the Impact of Public Land Management Practices on Bushfires in Victoria*, Parliamentary Paper No. 116 Season 2006–2008 (Melbourne: Government Printer, 2008)

[41] Victoria. Bushfires Royal Commission et al., "Final Report, Summary," 35. Five per cent represented roughly 380,000 hectares of the 7.7 million hectares considered public land.

[42] Victoria. Bushfires Royal Commission et al., "Final Report, Volume II," 295.

[43] Roger Underwood (Bushfire Front Inc.), *The February 2009 Bushfires in Victoria: A Submission to the Royal Commission from the Bushfire Front Inc.*, March 2009, http://bit.ly/2mrdGJn.

[44] Victoria. Bushfires Royal Commission et al., "Final Report, Volume II," 281–282.

failing to include any maps of vegetation in its reports.[45] The Commission's failure to do this or utilise its considerable resources to engage in a more critical analysis of the differences between Victoria and south-western Australia is indicative of how Australia is still fire illiterate. As stated, in 1939 Judge Stretton did recommend more prescribed burning, but also (with far less available expertise or knowledge) cautioned against the adoption of it in mountain ash country.

The Commission's Final Report was concluded in July 2010, and the then-government decided to support all recommendations—including Recommendation 56.[46] Over the next few years, the Department of Environment, Land, Water and Planning (DELWP) ramped up its prescribed burning commitment higher than at any point in Victoria over the last two decades; however, it never came within 100,000 hectares of the 380,000-hectare target.[47] After internal and public criticism that the 5 per cent target was not "achievable, affordable or sustainable", in 2015 the Inspector-General for Emergency Management (IGEM) was asked to review the policy.[48] IGEM recommended the hectare-based target be replaced with a strategy based around risk reduction. It argued DELWP now possessed both greater capability to determine the value of burning different areas (at least partly through the aid of computer simulation program PHOENIX RapidFire) and a greater level of ecological knowledge around the effects of prescribed burning on fire-sensitive vegetation and fauna.[49] In 2016, DELWP introduced a risk reduction target which aimed to reduce "residual risk"; by 2017, this shift was complete.[50] A simple state-wide measure based on area burned had been replaced with a system where Victoria was divided into seven regions, allowing a

[45] Tom Griffiths, "From the Ashes," *Inside Story*, 12 October 2011, http://insidestory.org.au/from-the-ashes.

[46] Inspector-General for Emergency Management and State of Victoria, "Review of Performance Targets for Bushfire Fuel Management on Public Land" (Department of Justice and Regulation, 2015), 2.

[47] IGEM, "Review of Performance Targets," 15.

[48] Ibid., 2.

[49] Ibid.

[50] Inspector-General for Emergency Management and State of Victoria, "Annual Report: Implementation of Recommendations on Bushfire Fuel Management" (Department of Justice and Regulation, 2017).

complicated measure of risk to be more precisely targeted to each region based on simulation modelling to compare scenarios involving different levels of prescribed burning. The hectare-based target had lasted less than a decade.

There is no doubt that the 5 per cent policy had increased the amount of land prescribed burnt in Victoria; however, it has been argued it led to the wrong kind of burning happening in the wrong places.[51] The Department of Environment and Primary Industries estimated that in 2012–2013 less than 3 per cent of Victorian "bushfire risk" was located in the Murray Mallee area, yet 16.9 per cent of the area prescribed burnt occurred there; although the more populated areas around Melbourne accounted for 31 per cent of risk, just 1.6 per cent of the total area of Victoria prescribed burnt was in this section.[52] This was not just a missed opportunity for prevention; the Mallee has many faunal species reliant on long-unburnt vegetation and this particular burning actively threatened their habitat.[53] This perverse policy outcome had been warned against by the Royal Commission's expert panel.[54] As IGEM noted, the shift to a risk-based strategy represented "a shift in focus from activity to outcome".[55]

The new risk-based strategy's commitment to regionalised prescribed burning targets is welcome and will do much to reduce the incentivisation of burning for hectares as opposed to burning for need. However, there is cause for scepticism around whether the new strategy will improve upon the second major reason for the 5 per cent target: public accountability. A significant justification behind the Commission's decision to recommend the 5 per cent target was that while crude, it represented a clear and easily understandable measure which would satisfy the public's considerable concerns around fire management. Even during the IGEM

[51] For example, David Lindenmayer et al., eds., *Mountain Ash: Fire, Logging and the Future of Victoria's Giant Forests* (Acton: CSIRO Publishing, 2015), 31.

[52] Andrew Bennett, Dale Nimmo, and Michael Clarke, "Burnoff Policies Could Be Damaging Habitats for 100 Years," *The Conversation*, 8 August 2014, http://theconversation.com/burnoff-policies-could-be-damaging-habitats-for-100-years-30240.

[53] Simon J. Watson et al., "The Mallee Fire and Biodiversity Project," *Proceedings of the Royal Society of Victoria* 124, no. 1 (2012): 38–46.

[54] Victoria. Bushfires Royal Commission et al., *Transcript of Proceedings Monday 22 February 2010*, February 2010, pp. 15197–15198, http://bit.ly/2kr7XCJ.

[55] IGEM, "Review of Performance Targets," 6.

review (conducted without the high profile of a Royal Commission), many of the 127 public submissions reflected a desire for greater transparency and accountability around prescribed burning.[56] A 5 per cent hectare target is easily understandable. A reduction of risk from 85 per cent to 60 per cent using an opaque algorithm-driven number is not; indeed, the use of PHOENIX RapidFire (or other simulators) has a political role in "rhetorically depoliticising subjective decisions by individuals and institutions about planned burning".[57] The new strategy has more room for fire nuances and may thus prove more ecologically effective. However, in the absence of greater public fire literacy, it may struggle to find legitimacy and thus prove less politically effective.

External Knowledge of Indigenous Burning: Broad but Not Deep?

A significant strain of public debate around prescribed burning concerned Aboriginal burning practices as these may have relevance to contemporary bushfire management. Prior to colonisation, Aboriginal groups across the continent used fire for a variety of purposes, and since colonisation some Aboriginal groups continued or have reengaged with these practices.[58] Aboriginal burning practices incorporated a broad range of material uses, including hunting, clearing areas of dangerous snakes, protecting and encouraging the growth of edible plants such as gunyang (*Solanum vescum*), and protecting sacred sites or valuable resources.[59] However, the extent to which this pre-colonial burning impacted upon the vegetation distribution of Australia is deeply contested.[60] This is

[56] Ibid.

[57] Timothy Neale and Daniel May, "Bushfire Simulators and Analysis in Australia: Insights into an Emerging Sociotechnical Practice," *Environmental Hazards* 17, no. 3 (2018): 208.

[58] It is unclear when Aboriginal burning substantially ceased as a practice in Victoria, although Aboriginal fire knowledge was retained by many groups. See Pyne, *Burning Bush*.

[59] See Beth Gott, "Aboriginal Fire Management in South-Eastern Australia: Aims and Frequency," *Journal of Biogeography* 32, no. 7 (2005): 1203–1208; Bill Gammage, *The Biggest Estate On Earth: How Aborigines Made Australia* (Crows Nest: Allen & Unwin, 2011).

[60] See DMJS Bowman, "Tansley Review No. 101: The Impact of Aboriginal Landscape Burning on the Australian Biota," *New Phytologist* 140, no. 3 (1998): 385–410.

especially vexing for contemporary management; if Aboriginal Australians burned drier areas in a certain fashion of low-intensity burns, and colonisation stopped this burning from occurring, fuel may have accumulated and encouraged high-intensity bushfires, which in turn might encourage a different pattern of vegetation, further altering fire dynamics.

In both the immediate post-fire popular discussion and during the Commission, the relevance of pre-colonial Aboriginal burning in Victoria was disputed, though references to Aboriginal burning were generally light on detail and heavy on rhetoric. Environmentalists argued that Aboriginal knowledge in Victoria was lost and the evidence of their practices was unreliable, that Aboriginal people used "cool burns" that were not "broadscale" (as opposed to higher-intensity modern prescribed burns that cover large areas), and that the very existence of large mountain ash forests proved that in at least some regions fire must have been excluded for a long time.[61] In contrast, many of the submissions supporting greater prescribed burning referenced Aboriginal burning to support their proposals. Quite a few referenced the speech of Wurundjeri elder Joy Murphy at a widely broadcast memorial service in Melbourne where she argued that Black Saturday was a consequence of the neglect of country since colonisation and that the burned areas had been burned on a seven-year cycle.[62] Murphy was not intending to give a detailed policy proposal, but it is striking how her speech was treated as though it were. Others drew upon evidence from other parts of Australia rather than the areas directly affected on Black Saturday. The Australian Forest Growers, for instance, contrasted James Cook's observations of what would become Cooktown in Far North Queensland with the same area today to argue about general changes in fuel patterns across Australia,[63] while others referenced observations of bushland in Sydney from 1789.[64] As stated, the precise patterns of pre-colonial Aboriginal fire practices—and more relevantly, their implications for contemporary prescribed

[61] Upper Yarra & Dandenongs Environmental Council, *Submission*.

[62] See Nillumbik Ratepayers Association, *NRA Submission*, May 2009, http://bit.ly/2mrdGJn.

[63] Australian Forest Growers, *Submission to the 2009 Victorian Bushfires Royal Commission*, May 2009, http://bit.ly/2mrdGJn.

[64] Richard Clarke (Building Designers Association of NSW), *BDA NSW Submission to the 2009 Victorian Bushfires Royal Commission*, May 2009, http://bit.ly/2mrdGJn.

burning—are not known with precision for the whole continent, but it is known that Indigenous burning was not homogenous in either method or effect.[65] It may have been the case that Victorian Aboriginal people did burn the wet ash forests in such a way that contemporary prescribed burning might replicate it (or eventually recreate more desirable vegetation patterns previously maintained by Aboriginal burning), but the evidence publicly available at this stage is less clear than for other ecological communities, so we can see that public debate after Black Saturday was broad, but not deep.

Since Black Saturday the public awareness of Aboriginal burning in southern Australia has grown, but as with those quoting Aunty Murphy or referring to practices from North Queensland, this does not necessarily reflect a greater attentiveness to fire literacy. The publication and wide circulation of Indigenous author Bruce Pascoe's *Dark Emu: Black Seeds: Agriculture or Accident?* and non-Indigenous historian Bill Gammage's *The Biggest Estate on Earth: How Aborigines Made Australia* have greatly spurred public awareness.[66] Indeed, there is evidence to suggest that the Australian public's understanding of Aboriginal burning practices is "inextricably tied" to their understanding of Aboriginal history and culture as a whole.[67] If, however, public understanding continues to be shaped by universalising narratives such as those delivered by Gammage and Pascoe,[68] there is a danger that discussion of Aboriginal burning will broaden, but not deepen—that policy advocates will advocate for inappropriate policy measures based upon information not necessarily applicable to the area in question. Indigenous-led "cultural burning" schemes which seek to restore and extend Aboriginal burning practices are

---

[65] Suzanne M. Prober et al., "Ngadju Kala: Australian Aboriginal Fire Knowledge in the Great Western Woodlands," *Austral Ecology* 41, no. 7 (2016): 716–732; R. J. Fensham, "Aboriginal Fire Regimes in Queensland, Australia: Analysis of the Explorers' Record," *Journal of Biogeography* 24, no. 1 (1997): 11–22.

[66] Bruce Pascoe, *Dark Emu: Black Seeds: Agriculture or Accident?* (Broome: Magabala Books, 2014); Gammage, *The Biggest Estate on Earth.*

[67] Billy Griffiths and Lynette Russell, "What We Were Told: Responses to 65,000 Years of Aboriginal History," *Aboriginal History* 42 (2018).

[68] Grace Karskens, "Fire in the Forests? Exploring the Human-Ecological History of Australia's First Frontier," *Environment and History* 25, no. 3 (2019): 391–419.

proliferating across southern Australia, even on privately held land.[69] In future bushfire inquiries, the relevance of Aboriginal burning practices to contemporary fuel reduction will no longer be an abstract question—but will include lived experience.

The future holds other complications for prescribed burning and fire literacy in Victoria. Changing demographics are likely to increase bushfire vulnerabilities.[70] Similarly, the effects of anthropogenic climate change are likely to increase bushfire risk. Although the exact nature of changes is uncertain, the number of days with low fire danger conditions is expected to decrease, while the number of days with dangerous conditions is expected to increase.[71] This not only increases the dangers of high-intensity firestorms like Black Saturday, but also represents a shrinking of the annual window in which prescribed burns are currently accepted as possible. This window is already short, averaging perhaps ten days a year in high-risk parts of Victoria.[72] Worryingly, high-intensity firestorms in some ecological communities can represent positive feedback loops; a crown fire in some dry eucalypt forests increases the probability of future crown fires.[73] Nevertheless, policymakers and agencies should not over-attribute climate change. Much of what made Black Saturday so devastating was not unprecedented—or indeed unexpected.[74] Overly enthusiastic causal explanations that rely on climate change rob individuals and institutions of their agency in the face of vast atmospheric forces and may encourage despair rather than action.

---

[69] Federation of Victorian Traditional Owner Corporations et al., "The Victorian Traditional Owner Cultural Fire Strategy," 2019.

[70] Holly Foster et al., "Peri-Urban Melbourne in 2021: Changes and Implications for the Victorian Emergency Management Sector," *The Australian Journal of Emergency Management* 28, no. 3 (2013): 6–11.

[71] Geoffrey J. Cary et al., "Global Change and Fire Regimes in Australia," in *Flammable Australia: Fire Regimes, Biodiversity and Ecosystems in a Changing World*, ed. Ross A. Bradstock, A. Malcolm Gill, and Richard J. Williams (Collingwood: CSIRO Publishing, 2012), 149–170.

[72] State Government of Victoria, *Report of the Inquiry into the 2002–2003 Victorian Bushfires* (Melbourne: State Government of Victoria, 2003), 20.

[73] James W. Barker and Owen F. Price, "Positive Severity Feedback between Consecutive Fires in Dry Eucalypt Forests of Southern Australia," *Ecosphere* 9, no. 3 (March 2018).

[74] Griffiths, "We Have Still Not Lived Long Enough."

Conclusion: The Need for Greater Fire Literacy

It is clear that prescribed burning can be an effective land management technique when done well. Improper application of it can have detrimental ecological effects or increase the danger of bushfires (as indeed can the improper withholding of fire), but even the most sceptical critic of prescribed burning might find it helpful to reflect upon the toxicology formulation of Paracelsus: "The dose makes the poison".[75] The contested conception and short life of the 5 per cent target demonstrates the question should not be 'to burn or not to burn'. This is a reductive formula that does our diverse country and environmental heritage a disservice. Ideally, there would be consideration of many questions enabled by a fire literate public, including how to burn (e.g. what intensities to aim for), when to burn (e.g. what time of year), where to burn (e.g. targeted or broad area burning) and why to burn (e.g. to mitigate against ember attack or to provide suppression 'anchor points')—and equally, how to avoid burns, when not to burn and where not to burn. Different fuel types react differently to fire. The term which dominated public debate after Black Saturday—fuel—is flattening, unconsciously hiding the differences between jarrah and mountain ash forests.[76]

Indeed, it is striking just how much of the commentary and public debate around prescribed burning in the ashes of Black Saturday reflected a lack of fire literacy. Aside from the usual confusion between prescribed burning and 'back burning' (a fire lit as a firefighting technique to deny fuel from an advancing bushfire's probable path),[77] there was no consensus upon how to refer to different types of prescribed burning for fuel reduction. Thus environmental groups could be accused of opposing prescribed burning even while they defensibly claimed to be in favour (albeit with qualifications). One explanation for this is that there is no consensus

[75] A discussion of this adage can be found in Sarah A. Vogel, "From 'The Dose Makes the Poison' to 'The Timing Makes the Poison': Conceptualizing Risk in the Synthetic Age," *Environmental History* 13, no. 4 (2008): 667–673.

[76] Philip Zylstra, "Forests, Not Fuels" (Bushfire Management—Balancing the Risks, Canberra: NPA ACT, 2017), http://www.npaact.org.au/res/File/2017/Phil%20Zylstra%20Forests.pdf.

[77] Greg Sheridan, "Crisis Survived, We Must Quickly Apply the Lessons," *The Australian*, 12 February 2009.

upon how to refer to different types of prescribed burning. A prescribed burn in strategic strips immediately around settlements might be called a 'targeted' burn as opposed to 'broad area' burns which seek to reduce the chance of fires becoming uncontrollable before they approach settled areas. Yet, this nuance was not conveyed in the submissions or media debate. As productive as it would be to stop thinking of fuel as simply 'fuel', perhaps we can move beyond thinking of fire as 'fire'. Even the Commission failed to reflect this nuance, choosing instead to confine its cultural recommendations to simply advocate agencies replace the American term wildfire with bushfire.[78]

Whether we are "still settling Australia"[79] or "we still have not lived long enough",[80] non-Indigenous Australians are struggling with antipodean disasters. The shallow debate after Black Saturday and the short-lived policy response demonstrate this. The new risk-based strategy reflects a greater emphasis on the ecological and demographic factors of fire management, but also represents a more opaque and less crudely accountable policy than a simple hectare-based target. The growth in public awareness of Indigenous burning is promising, but if it remains broad rather than deep, it will fail to legitimate or inspire better policy. A richer and more nuanced public understanding and language of prescribed fire (and Australian fire in general) is desperately needed.

Acknowledgements I am grateful to Professor Tom Griffiths, Dr Jillian Schedneck and especially my co-contributors to this collection for their warm and generous advice in preparing this chapter. This research was supported by an Australian Government Research Training Program (RTP) Scholarship, ORCID: 0000-0002-4749-3637.

---

[78] Victoria. Bushfires Royal Commission et al., "Final Report, Summary," 35.

[79] Stephen Dovers, ed., *Environmental History and Policy: Still Settling Australia* (Melbourne: Oxford University Press, 2000).

[80] Griffiths, "We Have Still Not Lived Long Enough."

Decolonising Settler Hazardscapes of the Waipā: Māori and Pākehā Remembering of Flooding in the Waikato 1900–1950

Meg Parsons and Karen Fisher

This chapter examines how different social groups within settler colonial societies understood and responded to the perceived environmental risks within particular landscapes and waterscapes. We explore the divergent viewpoints held by Māori and Pākehā (European/white) in the Waipā River catchment of Aotearoa New Zealand's North Island about what constituted safe, productive, and moral landscapes and waterscapes (Fig. 1). Such differences were underpinned by important ontological and epistemological differences, which influenced how both groups imagined and engaged with physical, moral, and metaphysical risks. Accordingly, we situate our discussion of flooding within the Waipā catchment not as series of individual disasters but, rather, as events within the wider and ongoing disaster of colonisation for local Māori whanau (families), hapū (sub-tribes), and iwi (tribes). Pākehā settler-led actions to

---

M. Parsons (✉) • K. Fisher
School of Environment, University of Auckland, Auckland, New Zealand
e-mail: meg.parsons@auckland.ac.nz; k.fisher@auckland.ac.nz

S. McKinnon, M. Cook (eds.), *Disasters in Australia and New Zealand*,
https://doi.org/10.1007/978-981-15-4382-1_9

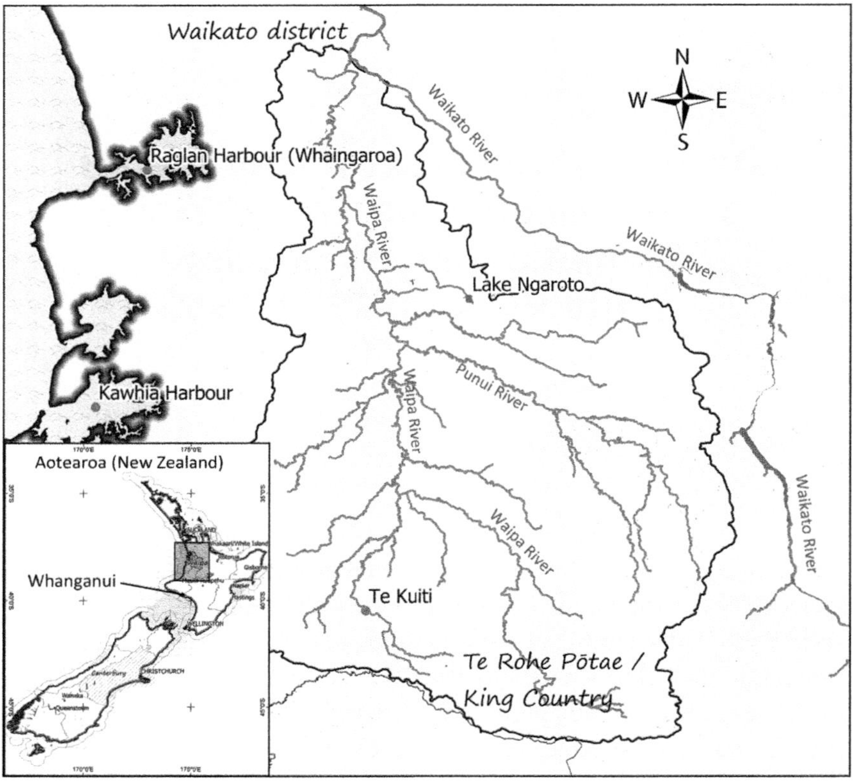

Fig. 1 Map of locations around the Waipā Valley region, Aotearoa (New Zealand)

radically remake Māori landscapes and waterscapes also transformed the hazardscapes of the region and resulted in increased vulnerability to flood and decreased capacities to respond to extremes amongst Māori due to increased poverty and socio-political marginalisation.

We adopt a historical geographical approach, drawing on archival sources, diaries, and oral histories, to explore how different groups perceived hazards within different landscapes and waterscapes comprising the Waipā River catchment from the 1860s until the 1940s. We focus particular attention on the period from 1900 until 1930 when widespread drainage and water engineering works took place as a way to erase the existing Māori landscapes and waterscapes of the Waipā and replace them with those of Pākehā. Our study provides an in-depth account of the diverse ways in which Pākehā settlers and Māori individuals, whanau, hapū, and iwi imagined, interacted with, and responded to environmental changes (both human-induced and natural variability) and how hazards and disasters are socially and culturally situated. More significantly, we seek to de-naturalise accounts of so-called ecological disasters by highlighting the links between colonialism, race, and power in historic and contemporary environmental crises (be it flooding, pollution, drought, soil degradation, and biodiversity loss) in Aotearoa New Zealand (hereafter Aotearoa).

Environmental risks are not simply biophysical or geological phenomena but also social and cultural and are, therefore, interwoven with different social imaginaries of landscapes and waterscapes. This chapter explores the Waikato region's history and the contrasting Māori and Pākehā viewpoints held about Waipā landscapes and waterscapes. Pākehā efforts to remake the existing indigenous landscapes and waterscapes of the Waipā increased vulnerability to flooding and diminished Māori capacities to maintain their economic, socio-cultural, and spiritual connections with their rohe (traditional lands and waters).

Situating Environmental Crises Within Historical Landscapes and Waterscapes

Both a landscape and a waterscape represent the power relations within a society as well as being an "instrument of cultural power".[1] Landscapes and waterscapes were and are ideological concepts through which people position themselves and their relationships with the land, water, biota, and nature. Over time, meanings are assigned to and inscribed onto particular landscapes and waterscapes, which come to define how people perceive, relate to, and interact with particular places and how they expect other people to behave. However, some meanings are privileged over

[1] Lise Saugeres, "The Cultural Representation of the Farming Landscape: Masculinity, Power and Nature," *Journal of Rural Studies* 18, no. 4 (2002): 375.

others, with meanings shaped by dominant ideologies (political, social, cultural, and economic). This includes how Māori and Pākehā cultures in the nineteenth and twentieth centuries interacted with nature and made assessments about what constituted a good or bad landscape or waterscape, and how they perceived (and experienced) environmental risks. Flooding within culturally situated landscapes and waterscapes of the Waipā was (and still is) co-constituted by human activities and reflective of struggles between different social groups (in our case study Māori and Pākehā), with inequitable power relations evident in people's differential capacities to access and use environmental resources, to manage hazards, as well as the quality and flow of water supplies.[2] These features shaped, rearticulated, and reproduced power inequities.

Background: Te Tiriti o Waitangi and the Invasion of the Waikato

While Te Tiriti o Waitangi (Treaty of Waitangi), signed by representatives of various Māori iwi and the British government in 1840, was supposed to guarantee Māori rights over (and possession of) their lands, waters, and other natural resource rights, colonial officials did not honour the terms of the Treaty.[3] The settler-led government actively sought to appropriate Māori land and limit the abilities of Māori to exercise rangatiratanga (chiefly authority) over natural resources using military, financial, and legal mechanisms. In the Waikato district, the centre of the Kīngitanga (King Movement) that opposed selling land to Pākehā, the appropriation of Māori land and waterways came first through military actions, with the British military invasion of the Waikato district between July 1863 and April 1864. The result of the Waikato War was the death of approximately 1700 people, as well as the destruction of the Waikato and Waipā Māori economies, and the raupatu (confiscation) of Māori lands

[2] Sue Jackson and Marcus Barber, "Historical and Contemporary Waterscapes of North Australia: Indigenous Attitudes to Dams and Water Diversions," *Water History* 8, no. 4 (2016): 386.

[3] James Belich, *Making Peoples: A History of the New Zealanders, from Polynesian Settlement to the End of the Nineteenth Century* (Auckland: Penguin Press, 1996).

(including those within the lower reaches of the Waipā and Waikato River catchments).[4] Iwi aligned with the Kīngitanga movement sought refuge with their kin Ngāti Maniapoto in the area south of the Pūniu River (a major tributary to the Waipā River), which became known as Te Rohe Pōtae by Māori (translated to English as the King Country, the name used by Pākehā).[5]

From 1864 until 1885, Te Rohe Pōtae was effectively off limits to Pākehā due to the aukati (boundary line) imposed by Ngāti Maniapoto in mid-1866.[6] In this way, Ngāti Maniapoto were able to exercise their mana (authority) over Te Rohe Pōtae land, waterways, and resources between 1864 and 1885. By the 1880s, the central government desperately wanted access to construct the central part of the North Island's Main Trunk Railway. Ngāti Maniapoto (and their Kīngitanga guests from other iwi) rejected initial government proposals for surveys and the railway.[7] After sustained negotiations between government officials, Kīngitanga leaders, and Ngāti Maniapoto chiefs between 1882 and 1885, survey of Te Rohe Pōtae and the construction of the railway line were allowed subject to the government agreeing to certain conditions including the protection of Ngāti Maniapoto's culturally important waterscapes.[8] However, railway construction facilitated the extension of instruments of settler colonial governance into Rohe Pōtae instigating Māori land alienation and loss of decision-making authority over resources. Settler initiatives could remake Māori landscapes and

---

[4] Vincent O'Malley, *The Great War for New Zealand: Waikato 1800–2000* (Wellington: Bridget Williams Books, 2016).

[5] Waikato-Tainui, Ngāti Apakura, and Ngāti Maniapoto all whakapapa (genealogy) trace their ancestry back to the Tainui waka (canoe), which brought their ancestors from Hawaiki. See Moepātu Borell and Robert Joseph, *Ngāti Apakura Te Iwi Ngāti Apakura Mana Motuake, Report for Ngāti Apakura Claimants and the Waitangi Tribunal* (Wellington: Crown Forestry Rental Trust, 2012).

[6] Hirini Moko Mead, *Tikanga Maori (Revised Edition): Living by Maori Values* (Wellington: Huia Publishers, 2016).

[7] Telegram from Lewis to Bryce, 10 January 1884, Folio 249, MA 13/93, Archives New Zealand, Wellington; "The Natives and Mr. Bryce's Promises," *Waikato Times*, 10 June 1884.

[8] Other promises included an amnesty on Māori "rebels", the right of iwi self-governance and the prohibition of liquor. The government only honoured its promise to grant the amnesty, all the other promises were not implemented. W. B. Otorohanga to Editor, *New Zealand Herald*, 30 June 1927.

waterscapes.[9] Government policies and actions were directly and indirectly designed to prevent Māori from accessing terrestrial and freshwater resources and control the flow of water.

Māori Waterscapes of the Waipā

Prior to the 1863–1864 war, the Waipā River and its tributaries were highly dynamic and resource-rich areas occupied by multiple iwi and hapū. The whakapapa (genealogies) and oral histories of different hapū (affiliated with the iwi of Ngāti Maniapoto, Waikato-Tainui, Ngāti Apakura, and Ngāti Hikairoa) recount that their rohe encompassed landscapes and waterscapes of the Waipā catchment, then and now.[10] These oral traditions and historical and archaeological studies demonstrate how waterscapes were significant places for Māori. Kainga (villages) and pā (fortified settlements) were situated adjacent to rivers, lakes, and wetlands. On higher ground, extensive kūmara (sweet potato) and taro cultivations were created. Wetlands were places of mahinga kai (food gathering) providing a diversity of flora and fauna including tuna (*Anguilla spp.*—short and long-tailed freshwater eels), fish and shellfish, berries, and bracken root, as well as vital sources of fibre used for clothing, rope, and construction of structures. Wetlands were also a place of refuge in times of conflict.

Oral traditions illustrate how the region's watercourses and wetlands were not static spaces, but changed with the consequence of variable water flow (drought and flood) as well as hapū-led management and modification. For example, Ngāti Apakura elders recount how their ancestors created Lake Ngāroto in the mid-eighteenth century through damming streams. Their ancestors built an island in the middle of the lake using large logs as the foundations (secured to the lake bed), over which they laid bundles of tree (mānuka, *Leptospermum scoparium*)

[9] Cathy Marr, *The Alienation of Māori Land in the Rohe Pōtae (Aotea Block) 1840–1920 Waitangi Tribunal, Rangahaua Whānui Series 8* (Wellington: Legislation Direct, 1996).

[10] Miria Tauriki et al., *Ngāti Maniapoto Mana Motuhake, Report for Ngāti Maniapoto Claimants and the Waitangi Tribunal* (Hamilton: Crown Forestry Rental Trust, 2012); Sean Ellison et al., *Wai 898 A99 Tainui Oral and Traditional Historical Report* (Wellington: Crown Forestry Rental Trust, 2012).

branches and reeds (raupō), and soil. A large mound was formed by the deposition of soil over generations of occupancy (home to roughly 200 people).[11] Archaeological studies support these oral traditions and identified there were at least five pā located around the lake edge as well as those constructed on the island, a place continuously occupied by Māori from its creation until residents were forced to flee during the 1863 invasion.[12] Wetlands like Ngāroto were significant resource extraction spaces, not limited to mahinga kai (food gathering sites).[13] In contrast to Pākehā perceptions of wetlands as unproductive and unhealthy spaces, for Māori, wetlands were carefully cultivated hydro-social waterscapes central to their iwi and hapū identity, health and well-being, and economic security.[14]

Within Māori ontologies, rivers and their tributaries are more than a course of running water; they form indivisible social-ecological-metaphysical wholes rather than individual systems. Unlike western scientific knowledge and dominant post-Enlightenment European intellectual thought, for Māori no clear divisions can be made between different components of freshwater systems. The Waipā catchment (be it the muddy wetlands, running rivers, and streams) comprised interconnected (culturally defined) waterscapes for local whanau, hapū, and iwi, which were (and are still) the embodiment of ancestors to whom iwi and hapū connect through whakapapa. They are home to flora, fauna, and hostile or friendly supernatural beings (taniwha) that take the form of eel-like creatures and guard the spiritual well-being of the waterways (and the hapū and iwi who are bound to them). A river is a taonga (something treasured and sacred), a more-than-human feminine being that possesses

[11] Pam Cromarty and Derek A. Scott, *A Directory of Wetlands in New Zealand* (Wellington: Department of Conservation, 1995).

[12] Moepātu Borell and Robert Joseph, *Ngāti Ti Apakura Te Iwi Ngāti Apakura Mana Motuake Report for Ngāti Apakura Claimants and the Waitangi Tribunal* (Wellington: Crown Forestry Rental Trust, 2012); Tauriki et al., *Ngāti Maniapoto Mana Motuhake*.

[13] R. D. Pick, "Waikato Swamp and Island Pa," *New Zealand Archaeological Association Newsletter* 11, no. 1 (1968): 30–35.

[14] Wilfred Shawcross, "The Ngaroto Site," *New Zealand Archaeological Association Newsletter* 11, no. 1 (1968): 2–29; R. D. Pick, "An Island Occupation Site on Lake Ngaroto on Pierce's Farm at the Northern End of the Lake," *The Journal of the Te Awamutu Historical Society* 2, no. 1 (1967): 19–21.

its own life force (mauri) and the provider of material and spiritual health and well-being.[15]

The Waipā is thus imbued with multiple meanings that traverse the economic, social, cultural, and metaphysical. To Ngāti Maniapoto, the Waipā River is an indivisible (female) entity commencing its journey from the mountain range at Pekepeke to its joining with the Waikato River and is home to their tribal kaitiaki (guardian) the taniwha known as Waiwaia. Ngāti Maniapoto oral histories emphasise, for example, how Waiwaia saved the lives of Maniapoto children who got into difficulty while swimming in the river. Waiwaia had "hundred lairs or hiding places along the course of the Waipa River", however, the ongoing reduction in estuarine water levels through settler and colonial government actions and engineering interventions from the late 1860s had negative effects on their kaitiaki, Waiwaia.[16] The centrality of these waterscapes to Ngāti Maniapoto iwi identity and well-being, and the indivisibility of social and ecological systems are encapsulated in one iwi members' oral history recounted to the Waitangi Tribunal:

> I turn now then gaze upon the hill that stands forth, [the wetlands of] Te Kawa, the place where eels were distributed to the thousands and to the lands of Ouruwhero. Let my gaze settle upon the running waters of Pūniu, the border between Maniapoto and Waikato. Waipā River the abode of the taniwha Waiwaia and Tuheitia, these are the waters of my elders.[17]

To local Māori floods were not disaster events but part of the functioning of the wider waterscape, with their tribal stories transmitted down through the generations through whakapapa, waiata (songs), and oral histories. A Ngāti Maniapoto waiata (song) highlights the ways in which people's understandings of the river and its functioning (including floods) centred on reciprocal relationships between the interwoven social, biophysical, and metaphysical worlds:

---

[15] George Barrett, *Wai 898 A109 Oral and Traditional History Volume Ngāti Tamainupō, Kōtara and Te Huaki* (Wellington: Crown Forestry Rental Trust, 2012).

[16] David Alexander, *An Overview of Selected Environmental and Resource Management Issues in Te Rohe Potae Inquiry District* (Wellington: Waitangi Tribunal, 2014), 286.

[17] Tauriki et al., *Ngāti Maniapoto Mana Motuhake*, 352.

> Like an atua [god] I wing my way into the heavens above! I gaze down! There below lies my river Waipā, cutting her way over the breast of my native land. My eyes brim with tears at the vision of splendour, 'tis the love of my river that meanders away. My eyes gaze intently upon the deep pools of the river they are myriad lairs of Waiwaia; the atua who gathers food for the people. The rocks of the river are an easy pillow for my head. The deep stretches of the river are a bed that rejuvenates my spirit and my body. I am sustained by the river, by taking the waters of the ancients, drawing the waters from the atua, by procuring the very water of life![18]

This stands in marked contrast to the dominant European lexicon that floods are destructive and morally "bad" events. These different views were clear soon after the 1863 invasion of the Waikato, when Pākehā began to experience and report on regular flood events (often during summer). On 25 December 1869, for instance, when a flood destroyed Māori cultivations along the Waikato and Waipā riverbanks (following a flood the same time the previous year), Pākehā remonstrated Māori for their "folly in persisting to cultivate [on] such low lying land". However, Māori declared floodwaters ensured the fertility of the soils and the "quality" of the low-lying land within their rohe was "so rich that they much prefer it to the hills, which were comparatively barren".[19]

According to Māori scholar Tom Roa, floods "were not a time of dread" for tāngata whenua (people of the land) but periods of "bounty since the repo (swamp) ... would be teeming with tuna and ducks ... from the flooded Waipā River".[20] Floods meant reduced time and energy for hapū spent fishing, hunting, and harvesting. Māori perceptions of their waterscapes—as places of abundance, safety, health, well-being, and identity—wherein flooding was an essential part of ensuring that the mauri (life force) was maintained which contrasted markedly with Pākehā perceptions of the Waipā waterscapes as risky and insecure spaces that required mitigative measures be taken as a matter of urgency.[21]

---

[18] Wikitōria Tāne, *Cultural Impact Assessment: An Assessment of Cultural Impacts of the Proposed Happy Valley Milk Ltd Dairy Factory on Redlands Road, Otorohanga* (Ōtorohanga: Nehenehenui Regional Management Committee, 2017).

[19] "Lower Waikato. Flood in the River.—Fire at the Canadian Flax Malls," *Daily Southern Cross*, 25 December 1869.

[20] Tāne, *Cultural Impact Assessment*, 47.

[21] Cromarty and Scott, *A Directory of Wetlands in New Zealand*.

Colonising Waterscapes: Pākehā Perceptions of Indigenous Environments

Prior to the 1863 invasion Pākehā perceived the Waikato region as an agriculturalist's paradise of fertile plains and lush cultivations but by the second half of the nineteenth century landscapes and waterscapes were framed as problematic and even hazardous spaces. As Pākehā sought to establish themselves within the newly acquired "backblocks" of Waikato and Rohe Potāe, the realities of "swampy" grounds was a source of ongoing anxiety, with waterscapes having imagined risks to safety, well-being, and viability of the settler body politic.[22]

To most Pākehā settlers and visitors, the Waipā catchment, with its stagnant waters, changeable flows, and expansive and seeping wetlands, did not resemble remembered (or imagined) modern European waterscapes of ubiquitous straightened, leveed, and regulated rivers. Wetlands, in particular, were seen to be the host of multiple (collective) environmental risks for which embryonic Pākehā farming communities were especially vulnerable.[23] The mere sight of wetlands, Pākehā travellers and local farmers reported, inspired in them feelings of dismay and melancholy. One traveller narrated his journey through the rivers of the Waipā and Whanganui in the early 1880s as one of the hardships through the "great forest wilderness", with the "whole place saturated with moisture for centuries".[24] The "swampy nature of the country and the truly dismal character of the whole surroundings" prompted him and his travelling companions to name the area the "Dismal Swamp".[25]

Settlers and government officials ascribed regular flooding events within the Waipā catchment to the persistence of wetlands, Māori

[22] For more about settler environmental anxieties, see James John Beattie, "Environmental Anxiety in New Zealand, 1850–1920: Settlers, Climate, Conservation, Health, Environment" (PhD diss., University of Otago, 2005).

[23] Geoff Park, "Swamps Which Might Doubtless Easily Be Drained: Swamp Drainage and Its Impact on the Indigenous," in *Environmental Histories of New Zealand*, ed. Eric Pawson and Tom Brooking (Auckland: Oxford University Press, 2002), 176–185.

[24] James Henry Kerry-Nicholls, *The King Country: Or, Explorations in New Zealand. A Narrative of 600 Miles of Travel Through Maoriland* (London: S. Low, Marston, Searle & Rivington, 1884), 265–266.

[25] Kerry-Nicholls, *The King Country*, 265–266.

harvesting practices (pā tuna or eel weirs), and the growth of trees along riparian zones. In response to perceived problems of flooding and wetlands nationally, the central government passed various legislations, beginning in 1875 with the *Napier Swamp Nuisance Act*, to facilitate state-authorised drainage of wetlands.[26] The 1893 *Land Drainage Act* allowed for the creation of drainage schemes with government subsidies. The *Land Drainage Act* 1904 established drainage boards, with members elected by local ratepayers. Given wide ranging powers, drainage boards could acquire private land (including Māori land), construct drainage works (despite local landowners' opposition), manage watercourses, and impose rates (taxes) on landholders. The legislation specifically included provisions that targeted Māori land and made Māori land subject to local government rates.

Between 1901 and 1926, 20 Commissions of Inquiry were undertaken by the central government to examine the nation's problematic rivers. The Inquiries, which concentrated on flooding, river "improvements" and drainage works were used to justify the decisions taken by drainage boards to systematically drain wetlands and re-engineer rivers, including the Waipā. The government's drainage policy, historian Geoff Park observes, was underpinned by four main ideas. Wetlands in their existing state were unproductive wastelands only valuable because of their development potential as fertile farmlands. Secondly, wetlands did not hold any scenic value and should not be preserved (unlike certain remnants of indigenous forests, birds, lakes, and mountains). Thirdly, legally, wetlands were future parcels of land where Māori entitlements (authority over and rights to access and use resources) were considered by both the courts and the government to transfer with land titles (once wetlands were "unwatered"). Fourthly, the value of transforming wetlands to farmland was considered of such national significance to requite governments (central and local governments) and individuals to intervene and fund it to ensure success of the process.[27]

---

[26] New Zealand Parliament, "Napier Swamp Nuisance Act 1875 (39 Victoriae 1875 No 4)" (1875); New Zealand Parliament, "Land Drainage Act" (1893); New Zealand Parliament, "Land Drainage Act 1904" (1904); New Zealand Government, "Rangitaiki Land Drainage Act" (1910).

[27] Park, "Swamps Which Might Doubtless Easily Be Drained: Swamp Drainage and Its Impact on the Indigenous"; Catherine Knight, *New Zealand's Rivers: An Environmental History* (Christchurch:

In the Waipā River catchment, drainage districts, overseen by at least 12 separate drainage boards, were quickly created in the first two decades of the twentieth century.[28] Since the majority of land in Rohe Pōtae (upper and middle Waipā catchment) was still Māori land (chiefly Ngāti Maniapoto), Māori landowners were liable to pay rates to their local drainage board, which often included multiple boards. The drainage boards were active in constructing canals and levees to control the flow of water and keep it separate from land, livestock, townships, and people. The *King Country Chronicle* stated in January 1932 that the residents of Te Kuiti township were largely unaware of the important work their local drainage board was undertaking on their behalf. The Mangapu Drainage Board, the reporter declared, was "working quietly but effectively in carrying out operations" which involved "clearing and straightening [of] the Mangaokewa and Mangapu Streams" (both tributaries of the Waipā). The board faced early challenges in its activities (due to the large portion of land still held by Māori) but as of 1932 was beginning to "show tangible results". Furthermore, once the drainage works were completed "the menace of flooding of hundreds of acres of first-class w[ould] be removed, and t[hose] areas w[ould] be suitable for intensive cultivation".[29]

Informed by colonial hydrological knowledge and technologies, as well as land use, spatial planning, tenure systems, and agricultural approaches imported from Britain, individual farmers and government officials positioned ongoing efforts to "tame" the waterscapes of the Waipā as of critical importance to the creation and maintenance of prosperous settler communities. However, in 1935 local government officials reported that, despite the Mangapu being "unwatered", the streams and rivers continued to be uncertain fixtures in the landscape. Indeed, after drainage works, the water in the Mangapu Stream was reported to "com[e] down very quickly" and "there [was] a big volume of water joining the Waipa [River]".[30] Thus, as the wetlands were drained, after heavy rain, with no wetlands or vegetation to absorb and slow down the process

Canterbury University Press, 2016), 180.

[28] "'Tua Tua Moana Swamp," *Waikato Times*, 25 May 1915.

[29] "Important Drainage Scheme," *King Country Chronicle*, 9 January 1932.

[30] Folder, C 58 395, BAAS A269 5113 Box 62, 96/434220. Archives New Zealand, Auckland.

into the Waipā River and its tributaries, there were ongoing issues with flooding.

Pākehā settlers believed floods were exacerbated by the continued use by Māori communities of pā tuna to catch tuna. Eel weirs, Pākehā protested to government officials, narrowed watercourses, contributed to sedimentation, created snags onto which debris could accumulate and impede the free flow of water, hindered efforts to drain the land and increased the risk of flooding by heightening floodwaters.[31] Complaints were made to local members of parliament and government officials, with settlers demanding the intensification of drainage works and the removal of eel weirs to mitigate flood risk. The Minister of Native Affairs, in response to a letter from the parliamentarian who represented the lower King Country, declared his Department was already well aware "that the weirs form a considerable bone of contention between the Natives and other residents in the locality".[32] However, he argued it was not a matter for central government to resolve but, rather, the responsibility of individual parties to reach legal agreements through the court system (with drainage boards able to remove eel weirs without consultation with or the permission of Māori).

Māori Contestation of Environmental Changes

Māori communities resisted, contested, and challenged this privileging of Pākehā settler colonial values and knowledge, Pākehā authority over the Waipā River and the consequences of this privileging in terms of resource rights, environmental degradation, and loss of biodiversity. Māori individuals and groups, for instance, regularly submitted petitions to members of parliament to request assistance. Most notably, throughout the twentieth century Māori throughout the Waipā catchment wrote to the Minister of Native Affairs and the Māori Members of Parliament to request that actions were taken to ensure Māori could retain possession of

[31] W. T. Jennings to Minister for Native Affairs, 24 April 1909, MA1 973, Archives New Zealand, Wellington.

[32] Minister of Native Affairs to W. T. Jennings, MA1 973, 4 May 1909, Archives New Zealand, Wellington.

their lands; alongside their demands that their unlawfully acquired lands were returned.[33] In 1920, for instance, Paiharae Tuhoro and whanau wrote to the Member of Parliament for Western Māori (Dr Maui Pomare) to protest about the local council diverting the river through their land and the construction of flood works. The letter of complaint reported that river engineering works reduced their capacities to access their land and waterways, and the amount the local government paid them in compensation was inadequate to make up for their economic and cultural losses.[34] Māori landowners had little power to challenge the governments' ability to take land under the *Public Works Act*; however, they were able to raise objections and seek better financial compensation than that offered by government. Māori objected to the acquisition of their land for drainage and flood controls works because of the negative physical impacts, and because such works contributed to even more Māori being dispossessed of their rohe.[35]

The records of Otorohanga flood control works provide detail about the actions that government officials took, but there is a noteworthy deficiency of Māori perspectives on drainage and flood control works in the early twentieth century. Consultation with the local community was limited. There was limited appreciation of Māori cultural values or economic values. Oral histories with local Māori recount how drainage works in the 1920s negatively affected whanau land. Piko Davis recounted: "You can fill the whole river bed but you can't alter the fact that it's a lower elevation than the rest of the paddock and of course it will automatically

[33] From 1868 until 1996 there were four Māori seats in the House of Representatives reserved for elected Māori members of parliament (MPs) (elected by Māori voters). From 1996 the number of seats increased to six Māori seats.

[34] District Engineer Murray to Under Secretary Public Works Department, 16 March 1920, W1 1211 48/110, 1, Archives New Zealand, Wellington.

[35] AJHR, *J1-Petition of Maniapoto, Raukawa, Tūwharetoa, and Whanganui Tribes* (Wellington: New Zealand Parliament, 1883); Department of Maori Affairs, "Petition No. 237-15 Raukete Te Hara and 27 Others," 4 September 1915, R22405777, AC1H, 16036, MA1, 1145, 1915/2835. Archives New Zealand, Wellington; Bamford & Brown to Kawa Drainage Board Clerk, 4 December 1909, BCDG A1492 Box 1, A16, Archives New Zealand, Auckland; E. J. Best to Mr Jennings, MP, 7 April 1909, MA1 973, Archives New Zealand, Wellington; Mr Fisher to Grace, 14 January 1909, MA1 973, Archives New Zealand, Wellington; AJHR, *G-06f Native Land Amendment and Native Land Claims Adjustment Act, 1922. Report and Recommendation on Petition No. 187/1922* (Wellington: New Zealand Parliament, 1923); Wahanui and 415 Others, "The Petition of the Kingites," *New Zealand Herald*, 16 July 1883.

flood".[36] Drainage operations, Māori residents argued, resulted in increased flood risk and provided government with justification for further engineering interventions to reduce damage to properties.

In submissions to government inquiries, Māori spoke out for more than a century about the consequences of drainage and flood control schemes in terms of the environmental degradation and negative impacts on their health and well-being.[37] Many of the flora and fauna species harvested by Māori from wetlands declined as a consequence of drainage and flood control works, dredging operations, gravel extraction, and the removal of vegetation. As a result, the traditional livelihoods of local Māori communities (which were subsistence-based and involved trading relationships with other groups) could not be sustained. Yet, while officials sometimes recognised that Māori complaints possessed merit, they continued to advocate that land development, wetlands drainage, and flood defences were of greater importance to the district and nation (socially, economically, and politically) than Māori livelihoods, cultural values, and well-being.[38]

One of the most significant damages and losses for local Māori was the destruction of tuna habitats and a decline in tuna numbers, closely linked to structural interventions to remake wetlands into grasslands and straighten meandering watercourses. Elders of Ngāti Apakura and Ngāti Maniapoto recalled how tuna provided more than food to their hapū, with the practices of harvesting and preserving serving to maintain important socio-cultural and spiritual connections between whanau, hapū, and iwi, their ancestors, and atua (gods). For Ngāti Maniapoto, in particular, tuna were and are still culturally significant animals. Their oral

[36] Michael Belgrave et al., *Te Rohe Pōtae Environmental and Wāhi Tahu Report* (Wellington: Crown Forestry Rental Trust, 2011), 202.

[37] Hone Te Anga v Kawa Drainage Board, 33 NZLR 1139 (High Court 1914); E. J. Best to Mr Jennings, MP, 7 April 1909, MA1 973, Archives New Zealand, Wellington.

[38] Throughout the late-nineteenth and twentieth centuries individuals and governments operated within Te Ao Pākehā (the Pākehā world), which contrasted with the Māori world (Te Ao Māori), and sought to apply their scientific knowledge and technologies to transform the perceived hazardous areas (wetlands, waterways, and marginal lands) into productive pastures, contributing to a 92 per cent decline in total area of wetlands (from pre-colonial era). See Meg Parsons, "Environmental Uncertainty and Muddy Blue Spaces: Health, History and Wetland Geographies of Aotearoa New Zealand," in *Blue Space, Health and Wellbeing: Hydrophilia Unbounded*, ed. Ronan Foley et al. (London: Routledge, 2019), 205–227.

histories record that their tūpana (ancestor) Maniapoto (the namesake of the iwi) had a pet tuna in Te Ana-Ureure.[39] Accordingly, the decline and loss of tuna from the Waipā catchment, which was interlinked with wider losses in native terrestrial and freshwater biodiversity, was felt by local Māori in terms of loss of mana (power and authority), mauri (life force), and relationships with each other (their whanau, hapū, iwi, and other iwi), their rohe, and the living (and non-living) beings. As Ngāti Maniapoto report:

> For many of the Ōtorohanga Māori community there remains a keen sense of loss relating to their land. With the loss of land came the destruction of pā, burial sites, living spaces, place names that recorded their tupuna oral history, ceremonial places, waterways, pā tuna, mahinga kai, horticultural kai, horticultural gardens, repo pātaka kai (wetland food baskets) destruction of native bush (used for pataka kai—food basket), and puna (fresh water springs) and waters that were used for all manner of purposes. These included, among other things water used for drinking, food gathering and preparation, cleaning, health promotion, spiritual cleansing, ceremonial rituals, birth rituals, rituals for preparation of deceased; most of which (puna) were destroyed in the draining of land for farming purposes earth moving for diversion of the Waipā River and construction of flood stop banks, construction of the Main Trunk Railway Line and subdivision developments of the Ōtorohanga Township for Pākehā settlement.[40]

As the aforementioned quote demonstrates, the Waipā River catchment provided generations of Ngāti Maniapoto with ample sources of kai (food), which included freshwater fisheries. The Waipā River allowed them to "to meet their obligations of manaakiatanga" (hospitality and kindness to others). An example of manaakiatanga is when visitors travel to marae (formal meeting place of a hapū) they are always provided with food by the tāngata whenua (local hapū); the hosts are obligated to provide local delicacies (such as tuna and shellfish) to their guests and if they fail to do so the hapū is perceived to lose some of their mana (prestige and social standing). The decline of freshwater resources and the problems it

[39] Tauriki et al., *Ngāti Maniapoto Mana Motuhake.*

[40] Tāne, *Cultural Impact Assessment*, 31–32.

creates in terms of the abilities of hapū and iwi to meet the cultural requirements of kaitiakitanga and manaakiatanga is a source of ongoing distress for Ngāti Maniapoto and other iwi.

This is a particularly important point when considering the ways in which disasters, changing environmental conditions, and disaster risk reduction strategies are understood and experienced by different groups of people. From a Pākehā worldview, traditional Māori foodstuffs (and broader Māori economic activities) could simply be replaced by practices more in line with "civilised" (European) modes of living. Yet, for Māori such activities were bound-up with their identities, values, and ways of life, and it was not simply a case of replacing one with the other. Even following the 1863 Waikato invasion and the 1885 opening of Rohe Pōtae, harvested foods were dietary stables for many whanau and allowed those with limited financial resources in the decades following colonisation to maintain healthy diets. In the present day, as numerous scholars observe, practices of gathering, preparing, and eating traditional foods with family, friends, hapū, and iwi remain a critical way that Māori maintain their cultural identities, reassert their links to their rohe, their hapū, and their ancestors, and allows them to learn and/or transfer mātauranga (knowledge) and practices across generations.[41] Through such activities, Māori enact the practices associated with their role as kaitiaki of their rohe.

Changing Vulnerabilities

Geographical and ecological scholarship highlights that hydrological engineering works often contributed to communities experiencing increased vulnerability to flood hazards. Scientists note how wetlands absorb excess water, which can slow the speed and reduce the height and forcefulness of floodwaters.[42] Thus, the loss of wetlands, coupled with vegetation change, sedimentation of riverbeds, and hydrological engi-

[41] Kura Paul-Burke et al., "Using Māori Knowledge to Assist Understandings and Management of Shellfish Populations in Ōhiwa Harbour, Aotearoa New Zealand," *New Zealand Journal of Marine and Freshwater Research* 52, no. 4 (2018): 542–556.

[42] Beverley R. Clarkson et al., "Wetland Ecosystem Services," in *Ecosystem Services in New Zealand: Conditions and Trend* (Lincoln: Manaaki Whenua Press, 2013), 195.

neering works, all resulted in alterations in the flow and behaviour of water. Actions to transform the Waipā waterscapes into landscapes inevitably increased the biophysical vulnerability of human communities to flooding events. In the Waipā River, the effects of seasonal small-scale flood events worsened as wetlands were lost and vegetation removed. The construction of hard structures (houses, factories, sheds, roads, railways, and so forth) on floodplains situated assets in vulnerable locations. This exacerbated the impacts of floods prompting public and political discussion about how to address flooding and ultimately resulted in the extension of existing approaches to flood risk management (most notably engineering works and clearance of riparian vegetation).[43]

Leeves designed to decrease flood risk served to accelerate Māori land loss. Interventions to drain wetlands and re-engineer the Waipā River, like elsewhere in Aotearoa, occurred irrespective of what Māori landowners desired, and negatively affected Māori livelihoods and well-being.[44] Since Pākehā residents were overwhelming strong advocates for drainage works and flood controls and comprised the majority of elected and non-elected local and central government officials, Pākehā priorities and approaches to resource and flood management took precedent over those of Māori. Furthermore, Māori experienced ongoing socio-economic deprivation as a consequence of the cumulative impacts of colonial violence, dispossession, and political marginalisation. Māori land holdings were frequently subjected to unwanted drainage works, and Māori were then charged taxes (local government rates) for those same drainage works, which they could not afford to pay. Failure to pay local government rates could result in legal proceedings (initiated by the local government) and if Māori could not pay their rates, the court could order their land be leased out or sold.[45]

While floods occupied Pākehā hazard imaginaries and mobilised an enormous amount of resources and effort to "control" the landscapes and waterscapes of the Waipā, the drainage works and structures implemented

---

[43] Letter from G T Wilkinson, ACGS 16211, J1 505/J, 1893/1155, 26 July 1893.

[44] Meg Parsons et al., "Disrupting Path Dependency: Making Room for Indigenous Knowledge in River Management," *Global Environmental Change* 56 (2019): 95–113.

[45] "Native Rates," *New Zealand Herald*, 30 June 1927.

to control the muddy blue spaces of the Waipā had far more disastrous effects on Māori than any flood could ever have. Māori lost not only their whenua as a consequence of drainage and flood control interventions, but also their waterscapes that possessed metaphysical and material importance to their health and well-being.[46]

Acknowledgements This research was supported by a grant from Marsden Fund, Royal Society of New Zealand. We would like to thank Dr Gretel Boswijk for her invaluable feedback on draft versions of this chapter.

[46] Tāne, *Cultural Impact Assessment*, 28.

Where the Wild Things Were

Deb Anderson

The stories a North Queenslander might tell you about Cyclone Yasi can swing from a world gone mad to a world plain "gone". As the Category-Five storm swept in from the Coral Sea on 2 February 2011, Emergency Services were *roaming the streets with loud hailers* in the small coastal town of Cardwell, *telling people they had to evacuate. But nobody told them where to go. And nobody told the single mums with no vehicle with their kids how to get there or where would be a safe place.*[1] The Bureau of Meteorology predicted wind gusts beyond 300 km/h while the radio, the television, *everything you turned on* had the emergency *sirens going all the time and it's in your face and the fear is ramped up but you can't not listen.* Even the old-timer

[1] The introduction is a collage of anecdotes, italicised to denote the representation of multiple voices, drawn from: Daryl Dickson and Geoff Moffatt, oral history interview with the author, 7 January 2014, transcript, "Bated Breath: The Lived Experience of Cyclone in a Climate Change World." Monash University.

D. Anderson (✉)
School of Media, Film and Journalism, Monash University,
Melbourne, VIC, Australia
e-mail: deb.anderson@monash.edu

S. McKinnon, M. Cook (eds.), *Disasters in Australia and New Zealand*,
https://doi.org/10.1007/978-981-15-4382-1_10

who'd seen out Larry (2006), Winifred (1986) and decades of other cyclones was telling the neighbours, *This time, I think we might need to run.*

Some did. Last-minute evacuees with *no intention of leaving* were suddenly throwing *silly things* in the car: *no clothes to mention, no socks that matched, hardly any shoes or anything that was going to help—a box of frozen prawns, a bottle of champagne, two mahogany gliders and a kingfisher!* Others couldn't. Girringun Aboriginal Corporation, a block back from the ocean swell, *had to stay open because people came there: forty lives there that night—with a newborn baby. Mind you,* further down Victoria Street *there was one guy in a wheelchair who was hiding behind the Community Centre and as the cyclone went 'round, he wheeled his wheelchair from one side to the other.* When the wind dropped and the eerily calm eye of that 600 km-wide cyclone crossed near Mission Beach that night, the local *police sergeant that shouldn't have been outside* was lured from his home-bunker by *the strangest light: it was white and the clouds were rolling over the edge of it like a waterfall … And in the middle of it all were all these birds, flying round and round. And then … suddenly … everything just went white.* (Farmers inland up the Great Dividing Range would wake to find *thousands and thousands and thousands of dead seabirds lying all over their paddocks.*)

In the stretch of time that trailed "monster killer" Cyclone Yasi,[2] evacuees would return through *almost medieval* towns dotted with military tents and lanterns to *a tangled mess* of housing, the *devastating* silence of an annihilated jungle and the *irrational* anger of the people who had stayed—now burdened with the memory, the *sense*, that *when they came out the world had gone … And more than anything,* these communities *would just like the world to go back to what it was before.*

Cautionary Tales of Catastrophe

Disasters can cut deep against the grain of ordinary human experience—and oral history can present a cautionary tale drawn from the lasting human struggle to come to terms with it. Oral historians have begun to

[2] Greg Stolz, Brian Williams, and Peter Michael, "Evacuations as 'Monster Killer' Cyclone Yasi Bears Down on Queensland," *The Daily Telegraph*, 2 February 2011, http://www.dailytelegraph.com.au/news/national/north-queensland-braces-for-cyclone-anthony-as-cyclone-yasi-brews-behind-it/story-e6freuzr-1225997788088.

reflect upon the distinctive value of this method of exploring the ways individuals and communities experience and make sense of disaster. As historian Stephen Sloan argued in the path-breaking international anthology, *Listening on the Edge: Oral History in the Aftermath of Crisis*, oral history can form a "response to the silences inherent in rescue, recovery and rebuilding" as the conventional media's mythologising of disaster unfolds.[3] In this vein, disasters are interpreted not as discrete violent events but rather as part of complex, ongoing processes of social and environmental change (media scholars Pantti et al. even suggest that disasters can be re-interpreted as "crises that have gone bad").[4] By taking account of the broader historical and cultural dimensions, research into disasters can engage deliberatively with social issues of vulnerability and precaution, structural issues of preparedness and response, and political issues of identity and responsibility. In turn, it can deliver warnings and ask reflexive questions about environmental concerns, about needs and desires in relation to loss and future sustainability. Crucially, it can also open up space for difference in disaster storytelling, for survivors grappling with an altered sense of the human place in the non-human, warming world.

The latter is the primary concern of this chapter. Here I draw from an ongoing Australian study into the lived experience of one "natural" disaster, severe tropical cyclone, which brings into relief cultural conceptions of weather, climate and change through the lens of oral history. A longitudinal collection of oral histories in cyclone-affected communities is under way in the Wet Tropics coastal region of Far North Queensland, in northeast Australia (2014-) (Fig. 1). An objective is to investigate the meaning of cyclone experience for local people and the mediated understanding of climatic risk—or, as the case may be, the core interpretive problems of anthropogenic climate change in their lived context.[5] Severe

---

[3] Stephen M. Sloan, "The Fabric of Crisis: Approaching the Heart of Oral History," in *Listening on the Edge: Oral History in the Aftermath of Crisis*, ed. Mark Cave and Stephen M. Sloan (Oxford: Oxford UP, 2014), 264.

[4] Mervi Pantti, Karin Wahl-Jorgensen, and Simon Cottle, *Disasters and the Media* (New York: Peter Lang, 2012), 17.

[5] Mike Hulme, *Why We Disagree About Climate Change: Understanding Controversy, Inaction and Opportunity* (Cambridge: Cambridge UP, 2009).

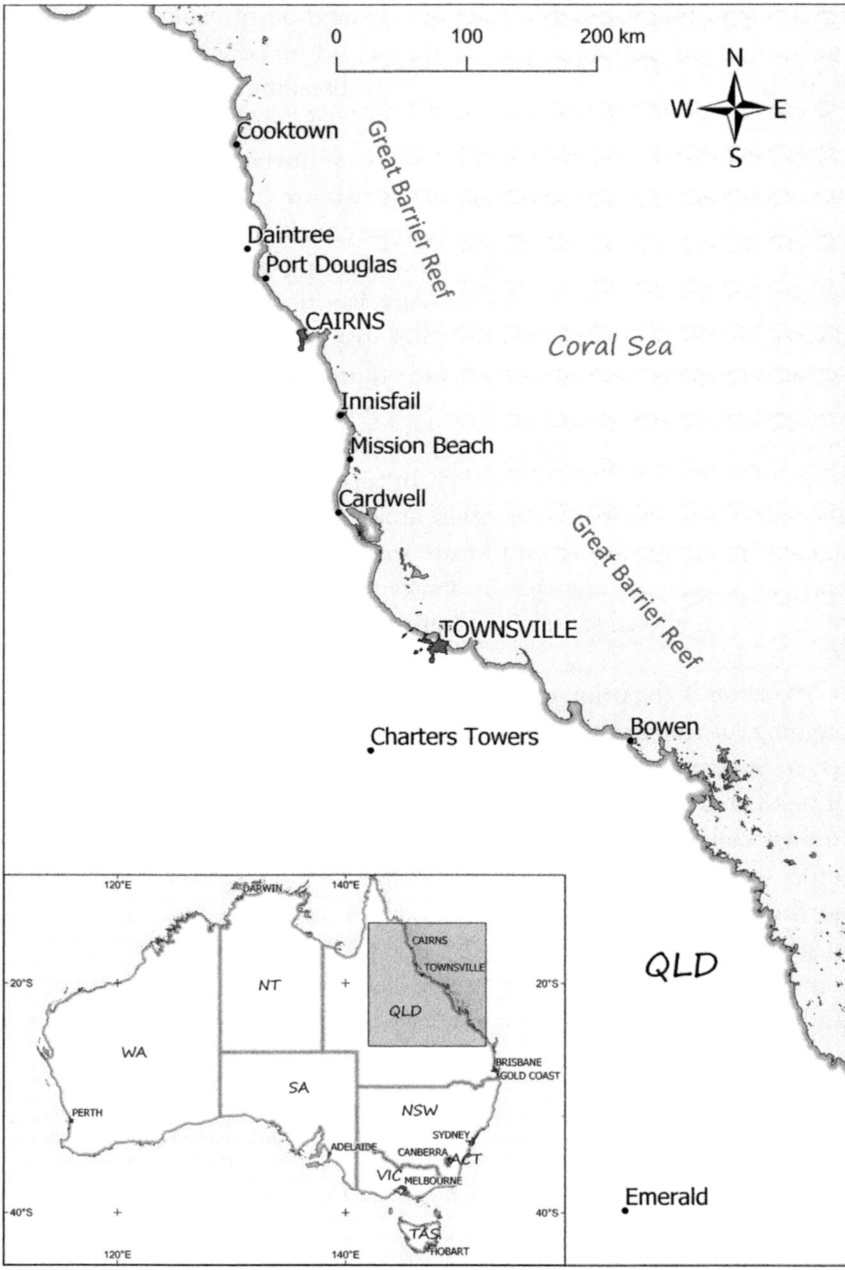

Fig. 1 Map of locations mentioned in Queensland

storms including Cyclone Yasi in Australia and Hurricane Katrina in the United States have sparked growing concerns over the impact of climate change on extreme weather events.[6] Australian scientists have been calling for urgent action—they named 2010–2020 the "critical decade"—to manage the risks of weather disasters and advance adaptation in light of climate change.[7] Yet, in 2011, fewer than half (45.6 per cent) of Australians accepted that global warming is caused by humans.[8]

This chapter argues for the value and application of oral history in shedding light on the ways people understand and struggle with cyclones as a core component of regional identity. In doing so, it homes in on the combined oral history of two North Queenslanders, that of Kennedy residents and wildlife conservationists Daryl Dickson and Geoff Moffatt. Theirs is among 20 oral history narratives recorded for the project so far, each of which presents a particular conception of disaster that shows different views of the relationship between humans and non-human nature. My choice to focus on one narrative is pragmatic. Insofar as their story is one of human and non-human environmental interconnection, "disaster" is experienced as the loss of "wild things"—a rich biodiversity encountered in everyday life—and in discourse of environmental (in)justice. I focus especially on segments of their story that engage with the social politics of belonging in cyclone country, illuminating the ways discourse on climate change sits at the intersection of biography, culture, politics and place. This choice also allows space for the richness of detail that oral history offers in contextualising, in a sense actualising, a cautionary tale on our imminent future.

[6] IPCC, *Managing the Risks of Extreme Events and Disasters to Advance Climate Change Adaptation: A Special Report of Working Groups I and II of the Intergovernmental Panel on Climate Change*, ed. C. B. Field, V. Barros, T. F. Stocker, D. Qin, D. J. Dokken, K. L. Ebi, M. D. Mastrandrea, K. J. Mach, G.-K. Plattner, S. K. Allen, M. Tignor, and P. M. Midgley (Cambridge: Cambridge UP, 2012); CSIRO and Bureau of Meteorology, *State of the Climate 2012: Second Report of the CSIRO and BOM* (Canberra: Australian Government, 2012).

[7] Will Steffen and Lesley Hughes, *The Critical Decade: Climate Science, Risks and Responses* (Australian Climate Commission, June 2013), https://www.climatecouncil.org.au/resources/critical-decade-risks-responses/.

[8] Zoe Leviston, Anne Leitch, Murni Greenhill, Rosemary Leonard, and Iain Walker, *Australians' Views of Climate Change* (Canberra: CSIRO Publishing, 2011).

Climate and Culture Close to Home

Issues of environment, climate and culture form something of a perfect storm in the deep north. This place is almost the opposite end of the country from my institutional base of Melbourne, which lies in Australia's far more densely populated southeast. Both Melburnians and north Queenslanders consider their weather "wild and unpredictable",[9] yet each has invested in their weather patterns the kind of meaning that can't be read in millimetres—or metres in the Tropics—or hectopascals. The fieldwork site is distant and culturally distinct, and in the context of climate denialism, it is a vital location for environmental historical research.

Cyclones themselves remain relatively under-researched in the fields of environmental history and cultural studies. In contrast to "The Three Horsemen"[10] of the driest inhabited continent—droughts, fire and flood—only a handful of studies have been conducted on the historical and cultural dimensions of cyclones in Australia.[11] Although cyclones have punctuated Australian regional histories, up to the time of Cyclone

[9] Heather Brown, "Tough Breed," *The Australian*, 4 February 2011, https://www.theaustralian.com.au/national-affairs/tough-breed/news-story/bfa791a9dc850dbeae0602b0ac795def; Brendan Casey, "Melbourne's Summer Brings Wild and Unpredictable Weather with Sun, Snow and Thunderstorms," *Herald Sun*, 5 December 2013, http://www.heraldsun.com.au/news/victoria/melbournes-summer-brings-wild-and-unpredictable-weather-with-sun-snow-and-thunderstorms/story-fni0fit3-1226776081717.

[10] Jay M. Arthur, *The Default Country: A Lexical Cartography of Twentieth-Century Australia* (Sydney: UNSW Press, 2003).

[11] Notable contributions include Bill Bunbury, *Cyclone Tracy: Picking up the Pieces* (South Fremantle: Fremantle Arts Centre Press, 1994); Peter Read, *Returning to Nothing: The Meaning of Lost Places* (Cambridge: Cambridge UP, 1996); Brad West and Philip L. Smith, "Natural Disasters and National Identity: Time, Space and Mythology," *Australian and New Zealand Journal of Sociology* 33, no. 2 (1997): 205–215; Alexis Wright, "Deep Weather," *Meanjin Quarterly* 70, no. 2 (2011): 70–82; Deborah Jordan, "Heeding the Warnings: 'Sucking up the Seas' in Vance Palmer's *Cyclone*," *eTropic*, no. 10 (2011): 20–31; Brad West, "Mythologising a Natural Disaster in Post-Industrial Australia: The Incorporation of Cyclone Tracy with Australian National Identity," *Journal of Australian Studies* 24, no. 66 (2000): 198–204; Stephen Muecke, "Hurricane Katrina and the Rhetoric of Natural Disasters," in *Fresh Water: New Perspectives on Water in Australia*, ed. Emily Potter, Alison Mackinnon, Stephen McKenzie, and Jennifer McKay (Carlton: Melbourne UP, 2007), 259–271; Alexis Wright, "Deep Weather," *Meanjin Quarterly* 70, no. 2 (2011): 70–82; Sophie Cunningham, *Warning: Cyclone Tracy* (Melbourne: The Text Publishing Company, 2014). Townsend this volume and PhD forthcoming; and more broadly Tim Sherratt, Tom Griffiths, and Libby Robin, eds., *A Change in the Weather: Climate and Culture in Australia* (Canberra: National Museum of Australia Press, 2005).

Tracy—which ripped through the Northern Territory capital city of Darwin on Christmas Day 1974 and was embraced as a national event—cyclones were characterised by "the conspicuous absence of nationalist discourse".[12] To this day, Australia's deadliest disaster, Cyclone Mahina (1899), has secured just a marginal place in the nation's collective memory (see Townsend's chapter "A Perfect Storm").

In North Queensland, local people (and the tourists that flock to the region) hear a lot about wild storms. Home to the planet's oldest rainforest, the Daintree, and its largest coral reef system, the Great Barrier Reef, the Far North is a region with more than one million visitors a year,[13] yet has a median household income lower than the national average.[14] This is also a region prone to tropical cyclones. On average, about eleven cyclones form in the Australian region each cyclone season, which is about 13 per cent of the global total, and about half of the total number become severe.[15] In Queensland, more than 200 cyclones have crossed the coast since 1858, when Commonwealth Bureau of Meteorology records on cyclone began; an average of 4.7 affects the Queensland Tropical Cyclone Warning Centre Area of Responsibility each year.[16] Forget "summer". The months of November to April up north are called "cyclone season".[17]

Given the locals must *live with* cyclone, I aimed to produce a historical account that centres not on disaster but on people and places, and which contextualises cyclone in a broader life story. That said, the oral histories have tended to focus on the experience of two recent, severe cyclones whose zones of impact overlapped. First came Category-Four Cyclone Larry, which crossed the coast near the sugar town of Innisfail on 20

---

[12] West, "Mythologising a Natural Disaster in Post-Industrial Australia," 198–204.

[13] James Cook University, "Campuses: About JCU," http://www.jcu.edu.au/about/campuses/ [accessed 10 January 2015].

[14] Department of Infrastructure and Transport Major Cities Unit, "Cairns," *State of Australian Cities 2013* (2013), https://www.infrastructure.gov.au/infrastructure/pab/soac/files/factsheets_2013/Cairns_Factsheet_FA.pdf.

[15] CSIRO and Bureau of Meteorology, *Climate Change in Australia—Information for Australia's Natural Resource Management Regions: Technical Report* (Canberra: CSIRO and Bureau of Meteorology, 2015); Bureau of Meteorology, "Tropical Cyclones in Queensland," http://www.bom.gov.au/cyclone/about/eastern.shtml [accessed 10 January 2015].

[16] Ibid.

[17] Jodie Davies, ed., *Cairns Region Storm and Cyclone Guide 2014* (Cairns: The Cairns Post Pty Ltd, 2014), 4.

March 2006 (remarkably, just one human death recorded). Five years later, the wider maximum-Category-Five Cyclone Yasi crossed about 50 km to the south, near Mission Beach (one death recorded then, too). This collection includes, for instance, stories of trying to escape the day before, or scrambling to open the window on one side of the house to ease a build-up of pressure inside your home-bunker, or praying to God as the wind kept getting stronger and the noise more terrifying, or hiding in a dilapidated shed and holding onto a pole while the shed was being blown to bits, or staying up all night wondering if family and friends were alive, or bathing yourself while washing the dishes in the pool for weeks afterwards.[18] Clearly, the approach and impact of Yasi were keenly felt in communities still recovering and rebuilding from Larry, when thousands of people lost their homes and livelihoods (my parents lost their house then too).[19]

A life-narrative approach enables the interviewer to gather far more than cyclone stories, however. Indeed "cyclone" forms a cultural site where historical stories of survival, both symbolic and literal, intersect.

The Call of the Wild

"Wild things" take both literal and symbolic form in the combined oral history of North Queensland wildlife conservationists Daryl Dickson and Geoff Moffatt, recorded in conversation in 2014. We were sweltering in the shade at Mungarru Lodge Sanctuary, their 3.5-hectare residence wedged within a fragmented maze of sugar cane fields, small-acreage rural blocks, seasonal and permanent creeks, state forest and national parks, in the lush Kennedy Valley about 11 km northwest of Cardwell. It is the kind of place where, typically, you can't see the sky for the trees. Three years on from Cyclone Yasi, whose impact on the region's rainforest areas was "almost akin to deforestation",[20] Dickson still grieves the loss of

[18] *Cyclone Yasi: Our Stories* (Cardwell, QLD: Cardwell and District Historical Society Inc., 2011).

[19] Deb Anderson, "Weathering the Storm: A Tale of Timing, Loss and Learning," *Traffic*, no. 9 (2007): 15–34.

[20] Laurence in Kerrin Binnie, "Cyclone Inflicts Long-term Damage to Rainforests," *ABC News*, 1 March 2011, updated 2 March 2011, https://www.abc.net.au/news/2011-03-01/cyclone-inflicts-long-term-damage-to-rainforests/1962988.

the "gentle, liquid" quality of light filtered through the forest canopy.[21] "It's like being put under a spotlight," she says, "where the light's far too bright and it's far too harsh, and it actually hurts your eyes."

Tellingly, Dickson and Moffatt speak of their home of two decades not in terms of property rights but instead human and non-human environmental interconnection: of the rich biodiversity encountered in everyday life. The pair, then in their 60s, introduces Mungarru Lodge Sanctuary as having a distinctive humane purpose. The sanctuary is situated on the banks of Meunga Creek, whose riparian corridor connects the Kirrama Range World Heritage Area, inland to the west, with the coastal Edmund Kennedy National Park and, to the east, the Great Barrier Reef marine reserve. "It's got platypus our end," Dickson begins. "It's got crocodiles across the highway."

Dickson: It's a fairly wild property—wilder now than it was three years ago because there's so much scrub down—but it's full of striped possums, was and will be again, sugar gliders, feather-tailed gliders, echidna, platypus, goannas, flying foxes, microbats … antechinus [marsupials that resemble mice].

Moffatt: I think … Daryl's counted 139 species of birds—

Dickson: Yes, it's so rich.

Moffatt: —and 19 species of mammals and endless amounts of reptiles, snakes.

Dickson: Oh, we get lovely big pythons.

Moffatt: We get pythons. I picked one up the other day that was about four metres long and took him for a little walk down the paddock before he ate my joey.

Dickson: Bandicoots …

Moffatt: Although it's 10 acres … I'd say there'd be only probably four of it which is actually flat enough to do anything with, so it's—

Dickson: So it's great for wildlife.

Moffatt: But it's a lifestyle block. It's not, it's not …

Dickson: It's a wildlife sanctuary. It's one of the Humane Society International's chain of wild places.

---

[21] All quotes from Dickson and Moffatt in this chapter are from: Dickson and Moffatt, "Bated Breath," 2014.

Of the 175 individual species of wildlife recorded in this particular "wild place", the gliding possums (or gliders) reserve a special status. These nocturnal, acrobatic marsupials possess a gliding membrane—a sheet of skin that stretches from forepaw to ankle. When extended, that membrane helps the possums "glide" distances of tens of metres, leaping from tree to tree.[22] The sanctuary's name is a nod to the fact that this is glider habitat and an acknowledgement of the long history and culture of the Traditional Custodians of the region: the Girramay People call gliding possums "mungarru".[23]

One species of glider is extra-special to the sanctuary, if partly for the wrong reason. "This is remnant mahogany glider habitat," says Dickson, "which is one of Australia's most endangered species and the one that we work with here." The sanctuary's existence thus also signifies a history of decline. The Wet Tropics Management Authority, jointly funded by Australian and Queensland governments and tasked with overseeing the Wet Tropics World Heritage Area (one of 200 areas where the Earth's biological wealth is most distinctive and rich), notes that the mahogany glider (*Petaurus gracilis*), once thought to be extinct, became endangered partly due to loss of habitat cleared for agriculture, residential sprawl and exotic pine plantations in the twentieth century. Other social and environmental factors to have impacted the elusive possum's natural habitat include inappropriate fire regimes and severe tropical cyclones, such as Yasi.[24]

Mungarru Lodge Sanctuary aimed to protect and regenerate that gliders' habitat. "Six years before the cyclone," says Dickson, "we had released the only glider bred in the captive breeding program that's ever been released into the wild." Named Stoney, he was the offspring of a glider they had sent to a breeding programme on the Gold Coast, in southeast Queensland. Four-year-old "Stoney was allowed to come back to us, and he partnered a wild glider that was injured". At Mungarru Lodge they

---

[22] Bush Heritage Australia, "Gliders," https://www.bushheritage.org.au/species/gliders [accessed 24 September 2019].

[23] Humane Society International, "About the Wildlife Land Trust," https://www.wildlifelandtrust.org.au/index.php/homepage/about [accessed 24 September 2019].

[24] Wet Tropics Management Authority, "Mahogany Gliders," https://www.wettropics.gov.au/mahogany-glider-krr.html [accessed 24 September 2019].

monitored this glider family and published a release diary about them online.[25] Stoney was "an absolute treasure"; he'd even graced the cover of *Wildlife Australia* magazine. "He taught us so much about the gliders. He never lost his human imprint, but he lived wild and he only came back to the built environment when he thought he could get an easy feed, or when he was in trouble."

Wild and Free?

Powerful emotions emerge as Dickson narrates her relationship to this place through work.[26] Intriguingly, the "work" of which Dickson speaks is the labour of care: of wildlife rescue, rehabilitation and release, most notably for the "graceful rope dancer" (her characterisation of the mahogany glider) but also for "any other wildlife that comes through the door" (such as the southern cassowary). Guiding this work is an ethic of "humane stewardship", which the Humane Society International describes as encouraging "members to practice sustainable and eco-friendly land management whilst preserving the valuable ecosystems and native species on their land".[27] For Dickson, this is a "passion" for "respecting and loving and enjoying the things that we've got for free in this world—for free in a way, but they're going to cost us an awful lot if we don't learn to look after them".

That passion has informed Dickson's *other* vocation, her artwork, which she and Moffatt have turned into a business, with a studio at Mungarru Lodge. Dickson paints local fauna and flora wildlife in water-colour (among other methods). Moffatt frames the artwork for exhibition and sale. The titles of the artwork—from "*Bathed in Colour*" *Eucalyptus* to "*Gentle*" *Pied Imperial Pidgeon*—connote a sense of Dickson's "place attachment" in the Far North yet reflect aspects of a

[25] Daryl Dickson, "Return to Country," last updated December 2010, https://daryldicksonartist.files.wordpress.com/2016/10/stoney-s-diary.pdf [accessed 24 September 2019].

[26] Debbie Lee and Kathryn Newfont, eds., *The Land Speaks: New Voices at the Intersection of Oral and Environmental History* (Oxford: Oxford University Press, 2018).

[27] Humane Society International, "About the Wildlife Land Trust," 2017, https://www.wildlifelandtrust.org.au/index.php/homepage/about [accessed 24 September 2019].

childhood spent on the other side of the world. Until she was seven, her family lived in London. They were "very much city people", she says, but her mother "was always rescuing something" and both parents always "engaged you with wild things". She is quick to clarify that point: "Not that my parents were environmentalists, by any stretch of the imagination."

That city-country binary and the politicisation of environmental knowledge were a larger part of Dickson and Moffatt's story. Like their parents, they were city dwellers once. Dickson emigrated to the South Australian capital city of Adelaide with her family in 1961; Moffatt was born and raised in Woodville in the western suburbs of Adelaide. They met while working in the Adelaide telecommunications industry. Says Moffatt: "She was a bit of a wanderer [smiles]. I think I just kind of tagged along." They "wandered" north in 1993, and their recollection of that "tree change" relays as much about the community as it does the countryside. Dickson says they envisioned the move as an "escape from the city", but "we had no idea what we were getting ourselves into". On arrival, they had to "scramble through jungle" to reach the land that would become Mungarru Sanctuary—land the locals called "rubbish scrub".

Dickson: We were the first Land for Wildlife property in the Far North, I think …

Moffatt: But when we put our … signs up for Land for Wildlife, one of our friends came down here and she said, "Do you really want to advertise that you're greenies? … Make yourselves a target?"

Dickson: Oh, she was really panicked about it … And it's … been probably one of the biggest challenges of living here.

This politics of environment and belonging underscored their narration of another challenge of life in the Wet Tropics: cyclones. Dickson and Moffatt were among those who evacuated as Yasi approached on 2 February 2011. "I think we also felt torn about running from the place," Dickson recalls. "That we wouldn't be here to try and help … our local community here when it was over, but that was the choice that we made because we just thought that the risk was too high." They were the ones with the champers, prawns and wildlife in the car. "All I could think at the time was, 'Gather up the things that are going to make you smile when you find you've got absolutely nothing left'." There was "no way" they could bring Stoney, she says: "He was a wild animal, so we didn't know where he was anyway, and we figured he'd just have to survive if he could." They first tried, but failed, to convince a neighbouring couple with three kids to join the evacuation. Then they drove south to Townsville, "not looking either side" of the car, then inland to Charters Towers ("the whole of Townsville was in Charters Towers"), then further south to Emerald, ending up almost 800 km from home. Emerald was flooded "and there was a big sign on the motel saying, 'Strictly no animals'".

It's Not Easy Being Green

Australian governments have been far from leading in taking action on climate change. As Australian National University scientist Tony Eggleton laments: "In Queensland, government officials do not refer to climate change; it is climate variability."[28] My reference to Australia's "deep north" is an allusion to the Deep South of the United States and the cultural stereotype linked to a region and most concentrated at its geographic extreme. Queensland's recent political history offers evidence that the sentiment epitomised by Joh Bjelke-Petersen's parochialism and frequent promotion of the idea of the secession of Queensland, during his term as premier in the 1970s and early 1980s, lingers in the Far North "extreme" today. Arguably, such a view of Queensland as *that* Queensland (an "embarrassing national joke—simply the product of police corruption, tawdry sex, draconian regulations and incompetent politicians"[29]) has been spurred on by political opportunism. Pauline Hanson's ultra-conservative One Nation Party still rated an outside chance in Queensland's 2015 knife-edge election.[30] But this is also a place where

[28] Tony Eggleton, "It's Global Warming, Stupid: Perceiving Climate Change," *Australian Quarterly* 84, no. 1 (2013).

[29] Julianne Schultz, "Disruptive Influences," *Griffith Review*, no. 21 (2008): 11.

[30] Michael McKenna, "Queensland Election 2015: Pauline Hanson a Chance for Return," *The Australian*, 3 February 2015, http://www.theaustralian.com.au/national-affairs/state-politics/

even less-conservative politicians, such as former Labor Premier Anna Bligh, can get away with a social-Darwinian appeal to the notion of resilience. As she told reporters in the aftermath of the 2011 "summer of disaster", which included flooding across an area the size of France, and Cyclone Yasi: "We are Queenslanders. We're the people they breed tough north of the border. We're the ones they knock down and we get back up again."[31]

The political utility of this mythologising of identity, place and nature is fascinating. Not only does the stereotype signify the power of discourse of identity and belonging (ever up for a fight) as psychological first-aid in the aftermath of upheaval. It also restricts ownership of, and access to, the experiential story of tropical disaster. Retired Australian Army General Peter Cosgrove, who led the Queensland Government taskforce of rebuilding communities damaged by Larry, put it bluntly in 2006: "No-one from outside the area of Far North Queensland hit by Cyclone Larry on March 20 can truly appreciate what the people here went through."[32] Where Cosgrove's statement implies an idyll of community and rurality that researchers have critiqued,[33] powerful ideas on rural solidarity persist as a source of comfort and reassurance in the presence of decline or wake of disaster.[34]

In my research, interviews so far have examined such storylines, illuminating the value placed on lived experience—and the right to dispute what this disaster meant. Some residents in the Innisfail district were adamant that Cyclone Larry was worse than Yasi, for instance. The *Innisfail*

queensland-election-2015-pauline-hanson-a-chance-for-return/story-fnr8rfrw-1227205659795.

[31] Trent Dalton, "How Cancer Taught Anna Bligh to Stop and Smell the Roses," *The Weekend Australian*, 7 December 2013, http://www.theaustralian.com.au/life/weekend-australian-magazine/how-cancer-taught-anna-bligh-to-stop-and-smell-the-roses/story-e6frg8h6-1226776024958.

[32] Innisfail Advocate, *Category Five Cyclone Larry, March 20, 2006: Your Stories, Tales of the Tempest* (Innisfail: Innisfail Advocate, 2006), 16.

[33] Barbara Pini and Belinda Leach, eds., *Reshaping Gender and Class in Rural Spaces* (Burlington: Ashgate, 2011); Stewart Lockie and Lisa Bourke, eds., *Rurality Bites: The Social and Environmental Transformation of Rural Australia* (Annandale, NSW: Pluto Press, 2001); Bill Pritchard and Phil McManus, eds., *Land of Discontent: The Dynamics of Change in Rural and Regional Australia* (Sydney: UNSW Press, 2000); Ian Gray and Geoffrey Lawrence, *A Future for Regional Australia: Escaping Global Misfortune* (Cambridge: Cambridge UP, 2001).

[34] For example, ABC Open's review of *The True Spirit of Cyclone Yasi*, 2012; and *Taken by Storm: Stories from People Affected by Severe Tropical Cyclone Larry*, 2007.

Advocate regional newspaper also defied the Bureau of Meteorology on the topic, publishing a volume of residents' cyclone stories that upgraded Larry from Category Four to Five.[35] Some interviewees said Cyclone Tracy (1974) still held the title of "worst". One woman I interviewed for the project had been through Yasi, Larry *and* Tracy. She was seven-and-a-half months pregnant the night Tracy hit Darwin and was part of the post-disaster mass-evacuation of mostly women and children. On the plane to Sydney, she had "that feeling" that something was wrong with the baby.[36] Her baby's death was registered in the Sydney Royal Women's Hospital, not the official Tracy toll.

In Dickson and Moffatt's depiction of life in the Far North, tensions over environmental knowledge are evident, which we might consider through the lens of "work" and "nature". As historian Richard White has observed, for many environmentalists, the two concepts have tended to "operate within a larger culture that encourages a divorce between the two".[37] White examined the tensions between people who identify as environmentalists and those who work in the same places, such as loggers and miners, in his provocative essay, "Are You an Environmentalist or Do You Work for a Living?" He argued that environmentalists too often equated "productive work" in nature with concepts of control or destruction. Intriguingly, the same contradictions White identified in the American environmental consciousness 25 years ago emerge here, in reverse, in a story of life in Australia's "deep north". The "jobs versus environment" debate is pointed in Queensland, where the political narrative has in recent state elections see-sawed from blinkered developmentalism to heavy-handed environmental regulation and back again.[38] In this respect, another factor shaping understandings of work and climate is the contemporary experience of crisis in many parts of regional Australia: the

---

[35] See Innisfail Advocate, *Category Five Cyclone Larry*.

[36] Tanya Blake, "Bated Breath: The Lived Experience of Cyclone in a Climate Change World," Oral history with the author, 3 January 2014. Monash University.

[37] Richard White, "Are You an Environmentalist or Do You Work for a Living?" in *Uncommon Ground: Rethinking the Human Place in Nature*, ed. William Cronon (London: W. W. Norton, 1996), 171.

[38] Allan Dale, "Jobs Versus the Environment: The Debate Queensland Can End," *The Conversation*, 23 February 2015, https://theconversation.com/jobs-versus-the-environment-the-debate-queensland-can-end-37760.

reality of social decline.[39] In that context, the Far North, environmentalists have been accused of "locking up", "quarantining" or even "sterilising" land, in turn condemning local communities to a future without jobs, economic development and infrastructure.[40]

Dickson and Moffatt's discourse of work denotes connection with the land through work that grants protection to it, and the pair recognises the challenge of their environmental ethic to the anthropocentrism embedded in traditional Western thinking. It marked them as "different" in the local community; after two decades, some locals still think they're "a bit strange". Dickson says when Moffatt approached a local teenager recently about the noise he was making on his dirt-bike in the bush, the boy's father arrived and said:

> "We knew we should have burned you greenie bastards out way back … as we'd planned when you first arrived here." And although I know that it's easy to say words, you wonder about the resonance of some of those things.

A Voice in the Wilderness

Life in the aftermath of Cyclone Yasi challenged any ideal of "natural" justice that could be read into Dickson and Moffatt's work or the environmentalism behind it. "You hear the stories and you feel people's individual pain while the story's being told," says Dickson. "But it's not until you're the main part of the story that you realise just how long people carry these things—for a lifetime." They may never be free of what this disaster, this history, means to them.

She and Moffatt drove four days to get home after Yasi, finding Mungarru Lodge just "a tangled, fallen mess". The real loss, however, was in and of the forest:

---

[39] Ian Gray and Geoffrey Lawrence, *A Future for Regional Australia: Escaping Global Misfortune* (Cambridge: Cambridge UP, 2001); Lockie and Burke, *Rurality Bites*; Pritchard and McManus, *Land of Discontent*.

[40] For example, Gerhardt Pearson, "Baubles and Trinkets Won't Make up for Mining Loss," *The Weekend Australian*, 27 May 2019, https://www.theaustralian.com.au/commentary/baubles-and-trinkets-wont-make-up-for-mining-loss/news-story/ab534c689a023c8f027640d540477c7f.

> The house isn't where we live … So the house almost didn't matter. I mean that's easy to say when the house was sort of still there, but the forest was, oh, it was devastating. There were no birds … the place was silent.

But what broke "the shock of everything" was a wild animal:

> When we finally got back here, our neighbours were really good: they left us to cry, to get over the shock … I'd walked into the kitchen. I was just standing there, I think numb, and suddenly this little, wet, bedraggled thing turned up at my feet.
>
> So, Stoney had survived. He was blind in one eye, which he'd been before the cyclone, and he had breaks in his tail and a hole in his patagium … but there he was. He was alive.

There is no neat, happy ending to this story, however. There were estimated to be as few as 1500 to 2000 mahogany gliders remaining prior to Cyclone Yasi. In the immediate aftermath, the habitat for these gliders was "devastated". The "blessing" of Stoney's survival was short-lived. With no food or shelter left for wild animals, Stoney and his family had to rely on feed stations, which made them vulnerable to predators. "Then the Rufous owl arrived and started eating everything that we loved—which is, I mean, good for the owls; they needed a feed—but they took Stoney and the whole family." *This* memory of disaster was symbolic of an uncertain future for Moffatt and Dickson, who now "hate" the word "resilient" and were struggling with an altered sense of the human place in the non-human world. "For 20 years we've talked about a species that's on the brink," says Dickson. "And [now] I look at them and I think, 'How much resilience is in anything?'" Indeed, the stories people tell about natural disasters not only reflect historical and cultural context but also, as cultural scholar Stephen Muecke has argued, organise human responses in the medium to long-term.[41] "Before the cyclone, if anybody said, 'Do you plan on leaving here?' I would have said, 'No way … because … this place is like finding where you belong'. [She looks to Moffatt] But it's changed, hasn't it?"

---

[41] Muecke, "Hurricane Katrina and the Rhetoric of Natural Disasters," 260.

Research shows that once climate change becomes "something people can feel, see or experience close to home", it becomes all the more meaningful[42]—and this research raises questions of the ways that weather disasters be viewed as climate-change-induced "compounded crises".[43] Three years on from Yasi, the mahogany glider's habitat was showing signs of recovery with shoot and leaf growth, flowering and limited fruiting recorded.[44] Yet scales of meaning were still unfolding for Dickson and Moffatt. For them, having experienced disaster, the dire warnings about extreme weather, climatic risk and the relationship between the two were accompanied by a sense of lost innocence about such warnings. Moreover, insofar as moments of disaster, as historian Stephen Sloan wrote, "can offer an environment when the larger weaknesses or strengths of a society are quite visible", here oral history sheds light on political, cultural and environmental realities that may be amplified under conditions of climate change.[45] For Dickson and Moffatt, these realities were no longer simply "sources of anxiety" woven into the fabric of everyday life.[46]

That said, the experience of Yasi and the cautionary tale it presents does reinforce a transpersonal ecology for this pair, where the memory of disaster is written into the future[47] and implications of human relationships with ecologies are better understood. This in itself forms a bid to make sense of the senseless. If and when North Queenslanders experience more of these intense cyclones in a warming world, says Dickson, then disaster relief funding will fall short: "The Government's not going to

[42] Sally Bingham, "Climate Change: A Moral Issue," in *Creating a Climate for Change: Communicating Climate Change and Facilitating Social Change*, ed. Susanne Moser and Lisa Dilling (Cambridge: Cambridge University Press, 2007), 163.

[43] Eva-Karin Olsson, "Responsibility Framing in a 'Climate Change Induced' Compounded Crisis: Facing Tragic Choices in the Murray–Darling Basin," *Environmental Hazards* 8, no. 3 (2009): 226–240.

[44] Queensland Department of Environment and Science, "Mahogany Glider," 2017, https://environment.des.qld.gov.au/wildlife/threatened-species/endangered/mahogony-glider#toc-3.

[45] Sloan, "The Fabric of Crisis: Approaching the Heart of Oral History," 265.

[46] Frederick Buell, *From Apocalypse to Way of Life: Environmental Crisis in the American Century* (London and New York: Routledge, 2003).

[47] Tom Griffiths, "Fire," in *Living with Fire: People, Nature and History in Steels Creek*, Christine Hansen and Tom Griffiths (Melbourne: CSIRO Publishing, 2012), 69–94.

have the resources to keep on pouring millions into making us all feel better." The answer, she says, lies in the promise of returning to living "closer to nature". We must teach our children to respect "wild things" (and where they are), to reconnect with the non-human world, and "how you make things out of nothing" in a world plain gone.

Index

A

Alexandra, 49
Alligator Creek, 102
All Souls St Bartholomew Quetta Memorial Church, 125
Annandale, 107
Aotearoa New Zealand, 159
Aplin's Weir, 107
Ash Wednesday, 8
Australian Capital Territory (ACT), 59

B

Back burning, 157
Barrow Point, 133
Bathurst Bay, 121
Bell, Charles Napier, 106
Black Friday bushfires, 144
Black River, 102, 113
Black Saturday, 8, 14, 21, 41, 140
Black Weir, 111
Bligh, Anna, 82, 87, 192
Bluewater, 113
Blue Water Creek, 102
Bohle, 101
Bowen, 104
Brazier, Felix Howard, 108
Brisbane, 80, 102
Brisbane River, 86, 102
Bushfires, 1, 21, 41, 60, 91, 139

C

Canberra, 59, 65
Cape Melville, 119
Cape Melville Range, 134
Cape York Peninsula, 119
Cardwell, 104
Christchurch, 9, 90
Cleveland Bay, 104
Climate change, 2, 183

S. McKinnon, M. Cook (eds.), *Disasters in Australia and New Zealand*,
https://doi.org/10.1007/978-981-15-4382-1

Colonialism, 161
Commemoration, 60
Cooktown, 154
Cosgrove, Peter, 192
Country Fire Authority (CFA), 41
Cyclone, 183
Cyclone Bola, 5
Cyclone *Debbie*, 136
Cyclone Larry, 185
Cyclone *Mahina*, 121, 185
Cyclone *Sid*, 113
Cyclone Tracy, 8, 90, 123, 184–185
Cyclone Yasi, 8, 179
Cyclones Larry, 114
Cyclones Yasi, 114
Cyclones, *Nachon*, 129

D
Daintree, 185
Darwin, 90, 123
Donnellys Creek, 48
Doogan, Maria, 64
Drought, 8, 23, 99
Dwellingup, 143

E
Earthquakes, 5, 79
Ecology, 196
Environmental history, 11

F
Far North Queensland, 181
Flinders Island, 134
Flood, 42, 80, 87, 99, 159
Flowerdale, 56

G
Gippsland, 42
Girramay People, 188
Great Barrier Reef, 185

H
Hanson, Pauline, 191
Heatley, 114
Hermit Park, 107
Hill, Jenny, 99
Historiography, 9–13
Holthouse, Hector, 123
Hurricane Katrina, 183

I
Indigenous, 13
Indigenous land management, 22

K
Kelso, 101
Kennedy, 183
Kennedy Valley, 186
Kilmore East, 56
Kinglake, 51

L
Lake Glenmaggie, 48
Landslides, 5

M
Magnetic Island, 113
Mahina, 4
Māori, 10, 14, 159

Marysville, 46
Melbourne, 42, 152
Memorial, 60, 125
Memories, 12, 32, 42, 61, 109, 123, 196
Mission Beach, 180
Mount Disappointment, 56
Mount Louisa, 110
Mount Stuart, 110
Mt Spec, 111
Murray Mallee, 152
Museums Victoria, 23

N
Newcastle, 79
Newman, Campbell, 95
New South Wales (NSW), 1
Night of Noah, 113
Nimmo, William HR, 110
1939 Black Friday, 141
1939 Royal Commission, 144
1961 Rodger Royal Commission, 143
Ninian Bay, 136
Northern Territory (NT), 3
North Island, 5

O
Oral history, 12, 23, 180

P
Pākehā, 159
Paluma Range, 111
Pearling disaster of 1899, 121
Poetry Tree, 28
Port Denison, 104
Prescribed burning, 140
Princess Charlotte Bay, 119

Q
Queensland, 1
Quinlan, Ted, 70

R
Race, 161
Recovery, 9, 23, 27, 60
Reinhold, William James, 110
Resilience, 13, 14, 29, 192
Ross Creek, 104
Ross River, 99
Ross River Dam, 101

S
Shannon River, 143
Somerset Dam, 87
South Island, 5
Stanhope, Jon, 59
"Stay or go" policy, 144
Strathewen, 25
Stretton, Leonard (Judge), 21, 141
Stuart, 111
Sydney, 154

T
Te Tiriti o Waitangi (Treaty of Waitangi), 162
Thursday Island, 119

Townsville, 99
Tsunami, 5
2009 Victorian Bushfires Royal
Commission, 144

V

Victoria, 3, 21, 42
Volcanic eruptions, 5
Vulnerability, 7, 14, 176

W

Waipā River, 159
Western Australia (WA), 3
White Island, 5
Whittingham, Herbert E., 123
Wivenhoe Dam, 80

Y

Yasi, 4

Printed by Printforce, the Netherlands